Reviews of Environmental Contamination and Toxicology

VOLUME 107

Reviews of Environmental Contamination and Toxicology

Continuation of Residue Reviews

United States Environmental Protection Agency
Office of Drinking Water Health Advisories

Editor
George W. Ware

VOLUME 107

Springer-Verlag
New York Berlin Heidelberg
London Paris Tokyo

New York 175 Fifth Avenue, New York, N Y 10010, USA
Heidelberg 6900 Heidelberg 1, Postfach 105 280, West Germany

Library of Congress Catalog Card Number 62-18595

ISSN 0179-5953

ISBN-13 978-1-4684-7085-7 e-ISBN-13 978-1-4684-7083-3
DOI. 10.1007/978-1-4684-7083-3

Foreword

Global attention in scientific, industrial, and governmental communities to traces of toxic chemicals in foodstuffs and in both abiotic and biotic environments has justified the present triumvirate of specialized publications in this field: comprehensive reviews, rapidly published progress reports, and archival documentations. These three publications are integrated and scheduled to provide in international communication the coherency essential for nonduplicative and current progress in a field as dynamic and complex as environmental contamination and toxicology. Until now there has been no journal or other publication series reserved exclusively for the diversified literature on "toxic" chemicals in our foods, our feeds, our geographical surroundings, our domestic animals, our wildlife, and ourselves. Around the world immense efforts and many talents have been mobilized to technical and other evaluations of natures, locales, magnitudes, fates, and toxicology of the persisting residues of these chemicals loosed upon the world. Among the sequelae of this broad new emphasis has been an inescapable need for an articulated set of authoritative publications where one could expect to find the latest important world literature produced by this emerging area of science together with documentation of pertinent ancillary legislation.

The research director and the legislative or administrative adviser do not have the time even to scan the large number of technical publications that might contain articles important to current responsibility; these individuals need the background provided by detailed reviews plus an assured awareness of newly developing information, all with minimum time for literature searching. Similarly, the scientist assigned or attracted to a new problem has the requirements of gleaning all literature pertinent to his task, publishing quickly new developments or important new experimental details to inform others of findings that might alter their own efforts, and eventually publishing all his supporting data and conclusions for archival purposes.

The end result of this concern over these chores and responsibilities and with uniform, encompassing, and timely publication outlets in the field of environmental contamination and toxicology is the Springer-Verlag (Heidelberg and New York) triumvirate:

Reviews of Environmental Contamination and Toxicology (Vol. 1 in 1962 as *Residue Reviews* through Vol. 97 in 1986) for basically detailed review articles concerned with any aspects of chemical contaminants, including pesticides, in the total environment with their toxicological considerations and consequences.

Bulletin of Environmental Contamination and Toxicology (Vol. 1 in 1966) for rapid publication of short reports of significant advances and discoveries in the fields of air, soil, water, and food contamination and pollution as well as methodology and other disciplines concerned with the introduction, presence, and effects of toxicants in the total environment.

Archives of Environmental Contamination and Toxicology (Vol. 1 in 1973) for important complete articles emphasizing and describing original experimental or theoretical research work pertaining to the scientific aspects of chemical contaminants in the environment.

Manuscripts for *Reviews* and the *Archives* are in identical formats and are subject to review, by workers in the field, for adequacy and value; manuscripts for the *Bulletin* are also reviewed but are published by photo-offset to provide the latest results without delay. The individual editors of these three publications comprise the joint Coordinating Board of Editors with referral within the Board of manuscripts submitted to one publication but deemed by major emphasis or length more suitable for one of the others.

Coordinating Board of Editors

Preface

Scope and Purpose of the Health Advisory Program

The United States Environmental Protection Agency (USEPA) Office of Drinking Water (ODW) Health Advisory Program was initiated to provide information and guidance to individuals or agencies concerned with potential risk from drinking water contaminants for which no national regulations currently exist. Health Advisories (HAs) are prepared for contaminants that meet two criteria: (1) the contaminant has the potential to cause adverse health effects in exposed humans, and (2) the contaminant is either known to occur or might reasonably be expected to occur in drinking water supplies. Each HA contains information on the nature of the adverse health effects associated with the contaminant and the concentrations of the contaminant that would not be anticipated to cause an adverse effect following various periods of exposure. In addition, the HA summarizes information on available analytical methods and treatment techniques for the contaminant.

History and Present Status

The program was begun in 1978, and guidance was issued for the first 20 contaminants in 1979. At that time, the concentrations judged to be safe were termed "Suggested-No-Adverse-Response Levels" (SNARLs). These guidance values were retitled Health Advisories in 1981. To date, the USEPA has issued 48 HAs in final form, covering a wide variety of inorganic and organic contaminants and one microbial contaminant (*Legionella*). In addition, USEPA has issues draft HAs on 50 pesticides that are being studied in the USEPA National Pesticide Survey, and is preparing additional HAs on various unregulated volatile organic chemicals, disinfectants and their by-products, and other inorganic contaminants.

Quality Assurance

Initial drafts of each HA undergo a series of thorough reviews before they are released to the public. The general technical content and the risk assessment values are reviewed by a group of independent expert scientists, an ODW Toxicological Review Panel, and any other USEPA offices with interest and expertise in the contaminant. The draft HAs are also distributed for review and comment by the public. Each HA is revised in response to criticisms and suggestions received during the review process before being released in final draft form. Each HA is periodically updated as significant new information becomes available that may impact the original conclusions or guidance values.

Acknowledgments

The development of each HA involves the participation of many individuals. The following members of the Health Advisory Program are acknowledged for their valuable contributions.

Michael B. Cook, Director, Office of Drinking Water

Joseph A. Cotruvo, Ph.D., Director, Criteria and Standards Division, Office of Drinking Water

Edward V. Ohanian, Ph.D., Chief, Health Effects Branch of Office of Drinking Water

Victor J. Kimm, Ph.D., Director, Office of Drinking Water (1975–1985)

William Lappenbusch, Ph.D., Chief, Health Effects Branch, Office of Drinking Water (1980–1985)

Jennifer Orme, M.S., Health Advisory Program Coordinator

Charles Abernathy, Ph.D.
Larry Anderson, Ph.D.
Ken Bailey, Ph.D.
Ambika Bathija, Ph.D.
Paul Berger, Ph.D.
Steve Clark, P.E.
Robert Cantilli, M.S.
Penelope Fenner-Crisp, Ph.D.
Julie Du, Ph.D.
Susan Goldhaber, M.P.H.
Maria Gomez-Taylor, Ph.D.
Krishan Khanna, Ph.D.
Peter Lassovsky, P.E.
Amal Mahfouz, Ph.D.
William Marcus, Ph.D.
Bruce Mintz, B.S.
James Murphy, Ph.D.
Myron Ottley, Ph.D.
Yogendra Patel, Ph.D.
Robert Vanderslice, Ph.D.

The members of the Health Advisory Program would like to acknowledge the assistance they have received from Kevin Gleason, Lori Gordon, Janet Normandy, John Glennon, and William Brattin of Life Systems, Inc., Jan Connery, Brana Lobel, and Heather Schroeder of Eastern Research Group, Inc., and Marlene Regelski of ECOS Management, Inc.

List of Acronyms and Abbreviations

AA	Atomic absorption
AADI	Adjusted Acceptable Daily Intake
ACGIH	American Conference of Governmental and Industrial Hygienists
ADI	Acceptable Daily Intake
ad lib	ad libitum
ae	acid equivalent
AWWA	American Waterworks Association
BUN	blood urea nitrogen
bw	body weight
CAG	USEPA Carcinogen Assessment Group
CAS	Chemical Abstracts Service
CDL	Company Document Library
CFR	Code of Federal Regulations
CHO	Chinese hamster ovary
Ci	Curie
CNS	central nervous system
DC	District of Columbia
DDT	dichlordiphenyltrichloroethane
DHHS	United States Department of Health and Human Services
dia	diameter
DNA	deoxyribonucleic acid
DWEL	Drinking Water Equivalent Level
EBCT	empty bed contact time
FAO	Food and Agricultural Organization of the United Nations
FDA	U.S. Food and Drug Administration
FSH	follicle-stimulating hormone
FWPCA	Federal Water Pollution Control Act
GAC	granular-activated carbon
GC	gas chromatography
GPT	glutamate-pyruvate transaminase

HA	Health Advisory
HCDD	hexachlorodibenzo-*p*-dioxin
HPLC	high-performance liquid chromatography
IARC	International Agency for Research on Cancer, Lyon, France
IgA	immunoglobulin A
IgG	immunoglobulin G
ip	intraperitoneal
IRDC	International Research and Development Corporation
iv	intravenous
LD_{50}	median lethal dose
LDH	lactic dehydrogenase
LH	luteinizing hormone
LOAEL	Lowest-Observed-Adverse-Effect Level
LOEL	Lowest-Observed-Effect Level
MAC	Maximum Allowable Concentration
MCL	Maximum Contamination Level
MLE	Maximum Likelihood Estimate
MRID	Microfiche Identification
MS	mass spectroscopy
NADPH	nicotine-adenine dinucleotide phosphate
NAS	National Academy of Sciences
NCI	National Cancer Institute, Bethesda, Maryland
NIOSH	National Institute for Occupational Safety and Health
NOAEL	No-Observed-Adverse-Effect Level
NOEL	No-Observed-Effect Level
NRC	National Research Council
NTIS	National Technical Information Service
NTP	National Toxicology Program
OCDD	octachlorodibenzo-*p*-dioxin
OCT	ornithine carbamoyltransferase
ODW	Office of Drinking Water
OPP	USEPA Office of Pesticide Programs
ORD	USEPA Office of Research and Development
OSHA	U.S. Occupational Safety and Health Administration
OTS	USEPA Office of Toxic Substances
PAC	powdered-activated carbon
ppm	parts per million
ppb	parts per billion
ppt	parts per trillion

RBC	red blood cell
Rfd	Reference Dose
RMCL	Recommended Maximum Contaminant Level
RNA	ribonucleic acid
RPAR	Rebuttal Presumption Against Registration
RSC	relative source contribution
S-9	high-speed supernatant (9,000-*g*)
SGOT	serum glutamic-oxalacetic transaminase
SGPT	serum glutamic-pyruvic transaminase
SNARL	Suggested-No-Adverse-Response Level
SOC	synthetic organic chemical
STEL	Short-Term Exposure Limit

Table of Contents

Acrylamide[a]

I. General Information

Acrylamide is also known as 2-propenamide, acrylic amide, acrylic acid amide, akrylamid, ethylene carboxamide, and propenoic acid amide. As a monomer, it is used in grouts and soil stabilizers. As a polymer, acrylamide is also used as a flocculant in drinking water and wastewater treatment plants, as an additive for enhanced oil recovery, as a fog dissipator, soil stabilizer, paper and paperboard strengthener, adhesive and binder component, and in metal coating, food packaging, photography, chromatography gel, electrophoresis gel, and dye applications. Its structure and properties are shown in Table 1.

The production of acrylamide in the U.S. in 1982 was estimated as 86 million lb (39.04 million kg) (USITC 1984). Acrylamide is used primarily in the production of polyacrylamide polymers and co-polymers. Approximately 1 million lb (454,000 kg) is used annually as a grouting agent (USEPA 1984).

Acrylamide monomer occurs as a contaminant in polyacrylamide. The monomer may be released to the environment during acrylamide production or use, or during the use of polyacrylamides. However, the major release results from use as a grout. Due to its low vapor pressure acrylamide is not released to air (USEPA 1984).

Acrylamide biodegrades in surface waters within a few days. Waters that are routinely contaminated by acrylamide degrade it even more readily (Brown and Rhead 1979). Hydrolysis of acrylamide to acrylic acid was reported, but this reaction is likely to be relatively slow (Brown and Rhead 1979; Brown et al. 1980b).

Acrylamide has not been studied in surveys of U.S. food and drinking water. Based on standards recommended by USEPA for polymers used in drinking water, levels of acrylamide monomer in drinking water reportedly have been as high as 0.5 μg/L (USEPA 1980). In one study in England, tap water levels of acrylamide were in the low-μg/L range (Brown and Rhead 1979). Acrylamide has not been reported in food but low levels may occur in some foods as a result of the use of polyacrylamides in food manufacture (USEPA 1984).

[a] Based on USEPA (1985a), except for Section I which is based on USEPA (1984)

Table 1. General information and properties of acrylamide

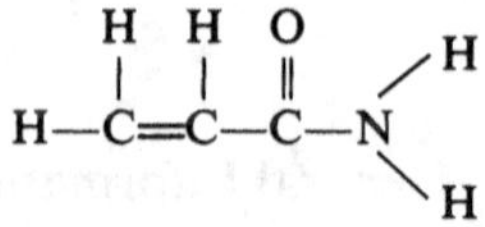

Property	Value
Chemical Abstracts Service (CAS) number	79-06-1
Chemical formula	C_3H_5NO
Molecular weight	71.08
Physical state (room temperature)	White crystals
Boiling point (25 mm Hg)	125°C
Melting point	84.5°C
Density	1.1
Vapor pressure (25°C)	0.007 mm Hg
Specific gravity (30°C)	1.122 g/mL
Water solubility (30°C)	2,155 g/L
Log octanol/water partition coefficient	—
Odor threshold	—
Taste threshold	—
Air concentration conversion factor	1 ppm = 2.95 mg/m^3

Source: Windholz (1976); Verschueren (1983)

II. Pharmacokinetics

A. Absorption

Acrylamide administered orally to rats (10 mg/kg) was absorbed rapidly and completely from the gastrointestinal tract (Miller et al. 1982). When blood levels of acrylamide after intravenous (iv) or dermal administration in rats were compared, approximately 25% of either dose of 2 or 50 mg/kg was absorbed through the skin (Ramsey et al. 1984).

In another report, 26% of a 0.5% solution of acrylamide was absorbed through the skin of rats in 24 hr (Franz et al. 1985). An additional 35% was present in the skin and potentially available for absorption. With the use of excised skin preparations, 67% (54% absorbed and 13% present in skin after washing) was either absorbed or available for absorption.

B. Distribution

After administering by gavage to rats, the highest concentrations were found in red blood cells, with lower amounts in all other tissues examined (Ramsey et al. 1984). Hashimoto and Aldridge (1970) reported that acrylamide was bound covalently to proteins and other cellular macromolecules.

Acrylamide freely crossed the placenta in pregnant female rats, rabbits, dogs, and pigs (Edwards 1976; Ikeda et al. 1983) and was uniformly distributed throughout canine and pig fetal tissue (Ikeda et al. 1983). Autoradiographic studies revealed acrylamide was widely distributed in male and female mice after oral administration of 120 mg/kg. The fetuses of pregnant mice were uniformly labeled except for a concentration of acrylamide in fetal skin (Marlowe et al. 1986).

C. Metabolism

In rats, acrylamide is metabolized primarily by conjugation with cellular glutathione (Miller et al. 1982). The major metabolite ($>50\%$) is the mercapturic acid *N*-acetyl-*S*-(3-amino-3-oxypropyl) cysteine, which was detected in the urine of rats given acrylamide orally or intravenously (Miller et al. 1982; Ramsey et al. 1984). Another metabolite resembling cysteine-5-propionamide was tentatively identified (Dixit et al. 1982).

D. Excretion

In rats, acrylamide and its metabolites are excreted primarily in urine (Miller et al. 1982; Ramsey et al. 1984). More than 60% of a dose administered orally or iv appeared in urine of rats within 24 to 72 hr (Miller et al. 1982; Ramsey et al. 1984). Minor routes of elimination ($<6\%$) included feces (Miller et al. 1982) and release of the amide carbon as carbon dioxide after oxidation (Hashimoto and Aldridge 1970; Ramsey et al. 1984).

III. Health Effects

A. Humans

Acrylamide intoxication was reported in three adults and two children exposed through ingestion of drinking water contaminated with 400 ppm acrylamide (Igisu et al. 1975). All adults exhibited symptoms of widespread central and peripheral nervous system dysfunction. The children apparently consumed less water and were less severely affected.

Additional reports of human exposure described primarily dermal or inhalation exposure of workers. The predominant effects included dysfunction of the central or peripheral nervous system or both. Quantitative dose and duration of exposure usually were unavailable (Auld and Bedwell 1967; Davenport et al. 1976; Fullerton 1969; Garland and Patterson 1967; Kesson et al. 1977).

B. Animals

Evaluation of the toxicologic database for acrylamide indicated that it is a cumulative poison. When the total dose administered over either short or longer periods of time reaches 100 to 150 mg/kg, neuropathologic signs began to arise in many species (USEPA 1985a).

1. Short-Term Exposure. Acute oral LD_{50}s for rats, guinea pigs, and rabbits ranged from 150 to 180 mg/kg (McCollister et al. 1964). Acute oral LD_{50}s for mice ranged from 107 to 170 mg/kg (NIOSH 1976; Hashimoto et al. 1981). The acute oral LD_{50} for acrylamide in male F-344 rats was 202.5 mg/kg (range, 188.9 to 217.3 mg/kg) (Pryor et al. 1983).

Single doses administered at levels as low as 25 mg/kg significantly increased binding of the neurotransmitter [^{3}H]spiroperidol in rat brains (Agrawal et al. 1981). Single doses of 1 to 100 mg/kg administered via intraperitoneal (ip) injection caused significant inhibition of retrograde axonal transport in rats at 25 mg/kg or greater. Doses of 1, 5, or 15 mg/kg caused no inhibition of transport (Miller et al. 1983).

In cats given acrylamide in the diet at 20 mg/kg/d for 2 or 3 wk, weakness of hind limbs and general unsteadiness of the posterior half of the body developed and usually progressed to hind limb paralysis (Leswing and Ribelin 1969). Microscopically, the affected nerves exhibited myelin and axonal degeneration. In dogs given 5 mg/kg/d, ataxia and muscular weakness developed by day 21 of treatment; demyelination of nerves was evident after 60 d (Thomann et al. 1974).

Rats given acrylamide in their drinking water displayed hind limb splaying after 14 d of treatment at 30 mg/kg/d (Gorzinski et al. 1979). Microscopic changes in peripheral nerves were observed in animals given 10 and 30 mg/kg/d. A No-Observed-Adverse-Effect Level (NOAEL) of 3 mg/kg/d was identified. When monkeys were treated with an average oral dose of 7.1 mg/kg/d in fruit juice, signs of visual impairment developed after 28 d; ataxia and motor impairment occurred after 46 to 65 d of exposure (Merigan et al. 1982).

2. Long-Term Exposure. The adverse health effects of acrylamide appear to have resulted primarily from damage to central or peripheral nervous system tissue. Characteristic effects were weakness and ataxia of hind limbs which progressed to paralysis with continued exposure (McCollister et al. 1964; Pryor et al. 1983; Thomann et al. 1974).

Subacute (5 d/wk for 4 wk) and subchronic (5 d/wk for 15 wk) LD_{50}s for acrylamide were 32.0 mg/kg (25.8 to 38.2 mg/kg) and 17.0 mg/kg (15.3 to 18.7 mg/kg), respectively (Pryor et al. 1983). Acrylamide administered in drinking water to rats at levels of 1 mg/kg/d for 90 d caused no external signs of toxicity, but histologic evidence of neuropathy (axolemmal invaginations) was noted (Burek et al. 1980). The NOAEL in this study was determined to be 0.2 mg/kg/d. Ataxia developed in cats that received 1 mg/kg/d orally for 125 d (Kuperman 1958). Weakness of the hind limbs developed in cats given

0.7 mg/kg/d for 240 d; a NOAEL of 0.2 mg/kg/d was identified (McCollister et al. 1964).

3. Reproductive Effects. Mice, given acrylamide orally, 10.1 mg/kg/d for 8 to 10 wk, displayed testicular atrophy, significant reduction in testicular weight, and degeneration of the epithelial cells of the seminiferous tubules (Hashimoto et al. 1981).

4. Developmental Effects. Acrylamide administered by gavage at 20 mg/kg/d to pregnant rats on days 7 to 16 of gestation significantly reduced [^{3}H]spiroperidol binding in striatal tissue of 2-wk-old pups (Agrawal and Squibb 1981).

5. Mutagenicity. Acrylamide did not elicit mutagenic activity in the Ames test using *Salmonella typhimurium* strains TA98, TA100, TA1535, and TA1537, with or without microsomal activation (Bull et al. 1984a). In the hepatocyte primary culture deoxyribonucleic acid (DNA) repair test, it did not exert mutagenic effects (Miller and McQueen 1986). It did induce chromosomal breakage and aberrations in spermatogonia of mice exposed to 75 mg/kg/d in the diet for 2 or 3 wk (Shiraishi 1978).

In a dominant lethal study, male rats received acrylamide in drinking water at 0, 15, 30, or 60 mg/L (0, 1.5, 2.8, or 5.8 mg/kg/d) for 80 d (Smith et al. 1986). Males were mated to untreated females that were killed on day 14 of gestation. A significant increase in preimplantation loss was noted in females mated to males given 60 mg/L. Significant postimplantation loss was observed in females mated to males given doses of 30 or 60 mg/L. The authors concluded that acrylamide produces dominant lethality in male rats. This effect was noted at dosages in which no splaying of hind limbs was evident and in which significant histopathologic lesions of the sciatic nerve occurred as determined by light microscopy.

6. Carcinogenicity. Groups of male and female Fischer 344 rats received drinking water containing acrylamide monomer at 0, 0.01, 0.1, 0.5, or 2.0 mg/kg/d for 2 yr (Johnson et al. 1986). After 1 year, significant depression of body weight was observed in males given the highest dose. Distal neuropathy occurred in the peripheral nerves of animals in this group. Tumor incidence was not increased significantly in groups that received 0.01 or 0.1 mg/kg/d. Male rats given 0.5 mg/kg/d had significantly increased incidences of scrotal mesothelioma. Statistically significant increased incidences of tumors in the following tissues were determined in rats treated at 2.0 mg/kg: females—mammary gland (benign and malignant), central nervous system (malignant), thyroid gland follicular epithelium (benign and malignant), mouth (benign), uterus (malignant), and clitoral gland (benign); males—scrotal mesothelioma (malignant) and thyroid gland follicular epithelium (benign).

Female Swiss ICR mice given acrylamide orally at 5.4, 10.7, or 21.4 mg/kg/d for 2 wk had a dose-dependent increase in tumors induced by the phorbal ester 12-*O*-tetradecanoylphorbal-13-acetate (TPA) (2.5 μg/mouse, 3 times/wk for 20 wk) (Bull et al. 1984a). Male and female mice that received acrylamide orally or intraperitoneally at 2.7, 5.4, or 10.7 mg/kg for 8 wk showed statistically significant increases in the incidence of lung adenomas (Bull et al. 1984b). Acrylamide was more potent when administered by gavage than when given by systemic routes.

IV. Quantification of Toxicological Effects[a]

No adequate dose–response data representing oral exposure are available to develop short-term risk assessments. However, in view of substantial chemical disposition evidence showing that acrylamide is absorbed rapidly and completely by virtually any route of exposure, it is considered acceptable to use data derived from exposure via other routes.

A. One-Day Health Advisory

The results of Miller et al. (1983) are considered appropriate for use in calculating the One-Day Health Advisory (HA). Male Sprague–Dawley rats (five/dose) were injected intraperitoneally with a single dose of acrylamide (1 to 100 mg/kg), and the rate of retrograde axonal transport of iodinated nerve growth factor was measured. Significant inhibition of transport occurred at doses of 25 mg/kg or more, but no significant changes were seen at 15 mg/kg or less. A NOAEL of 15 mg/kg was identified. The One-Day HA for a 10-kg child can be calculated as follows:

$$\text{One-Day HA} = \frac{(15\ \text{mg/kg/d})(10\ \text{kg})}{(100)(1\ \text{L/d})} = 1.5\ \text{mg/L}\ (1{,}500\ \mu\text{g/L})$$

B. Ten-Day Health Advisory

The results of Gorzinski et al. (1979) are considered appropriate for use in calculation of the Ten-Day HA. Acrylamide was administered at levels of 0, 1, 3, 10, or 30 mg/kg/d in drinking water to male and female Fischer 344 rats for 21 consecutive d. Histologic examination of peripheral nerves with both light and electron microscopy determined that axonal degeneration and demyelinization occurred at 10 and 30 mg/kg/d; no significant changes were apparent at doses of 0, 1, or 3 mg/kg/d. A NOAEL of 3 mg/kg/d was identified. The Ten-Day HA for a 10-kg child is calculated as follows:

$$\text{Ten-Day HA} = \frac{(3\ \text{mg/kg/d})(10\ \text{kg})}{(100)(1\ \text{L/d})} = 0.3\ \text{mg/L}\ (300\ \mu\text{g/L})$$

[a] See Introduction for a discussion of methodology

C. Longer-Term Health Advisory

The results of Burek et al. (1980) are considered appropriate for deriving the Longer-Term HA. Acrylamide was administered in drinking water for 90 d to male and female CDF rats at 0, 0.05, 0.2, 1, 5, or 20 mg/kg/d. Electron microscopy revealed that animals given 1 mg/kg/d exhibited axolemmal invaginations of peripheral nerves. No significant alterations were observed at 0, 0.05, and 0.2 mg/kg/d. Thus, based on the most sensitive measure of toxicity (i.e., ultrastructural examination of peripheral motor nerves), 0.2 mg/kg/d was identified as the NOAEL. The Longer-Term HA for a 10-kg child is calculated as follows:

$$\text{Longer-Term HA} = \frac{(0.2\ \text{mg/kg/d})(10\ \text{kg})}{(100)(1\ \text{L/d})} = 0.02\ \text{mg/L}\ (20\ \mu\text{g/L})$$

The Longer-Term HA for a 70-kg adult is calculated as follows:

$$\text{Longer-Term HA} = \frac{(0.2\ \text{mg/kg/d})(70\ \text{kg})}{(100)(2\ \text{L/d})} = 0.07\ \text{mg/L}\ (70\ \mu\text{g/L})$$

D. Lifetime Health Advisory

Burek et al. (1980) is the most appropriate study from which to derive the DWEL. The experimental details are described in the Longer-Term HA section. An additional uncertainty factor of 10 is included to accommodate for use of a study of less than lifetime duration. From this study, a NOAEL of 0.2 mg/kg was identified.

The RfD and DWEL are calculated as follows:

Step 1: Determination of the Reference Dose (RfD)

$$\text{RfD} = \frac{(0.2\ \text{mg/kg/d})}{(1{,}000)} = 0.0002\ \text{mg/kg/d}$$

Step 2: Determination of the Drinking Water Equivalent Level (DWEL)

$$\text{DWEL} = \frac{(0.0002\ \text{mg/kg/d})(70\ \text{kg})}{(2\ \text{L/d})} = 0.007\ \text{mg/L}\ (7\ \mu\text{g/L})$$

Step 3: Determination of the Lifetime Health Advisory

A Lifetime Health Advisory is *not* recommended because acrylamide is classified as a probable human (Group B2) carcinogen (see *Evaluation of carcinogenic potential* below). For comparison purposes, the cancer risk associated with the DWEL was determined.

The estimated excess cancer risk associated with lifetime exposure to drinking water containing acrylamide at 7 μg/L is approximately 7×10^{-4}. This estimate represents the upper 95% confidence limit from extrapolations prepared by the USEPA Carcinogen Assessment Group (CAG) using the linearized, multistage model. The actual risk is unlikely to exceed this value.

E. Evaluation of Carcinogenic Potential

When the criteria described in USEPA guidelines for assessment of carcinogenic risk (USEPA 1986) are applied, acrylamide is classified in Group B2: probable human carcinogen. Group B2 contains substances with sufficient evidence of carcinogenicity in animals and inadequate evidence from human studies.

Studies in mice (Bull et al. 1984a,b) and rats (Johnson et al. 1986) showed that acrylamide has significant carcinogenic potential. Based on a drinking water study in rats (Johnson et al. 1986), CAG prepared a draft quantitative risk assessment of acrylamide exposure (USEPA 1985c). In this draft assessment, CAG derived several carcinogenic potency factors from different sets of dose–response data. CAG recommended, however, that the human cancer potency factor (q_1*) of 3.7 $(mg/kg/d)^{-1}$ (derived from combined data for the incidence of tumors of mammary glands, thyroid, and uterus in females) be used for estimation of the increased lifetime risk of human exposure to acrylamide. A 70-kg adult is assumed to ingest 2 L/d of water over a 70-yr lifetime, and therefore the concentrations of 1 μg/L, 0.1 μg/L, and 0.01 μg/L are associated with estimated excess cancer risk of 10^{-4}, 10^{-5}, and 10^{-6}. These estimates were made by the Office of Drinking Water (ODW).

V. Other Criteria, Guidance, and Standards

Polyacrylamide products used as coagulant aids in the treatment of drinking water should not have a residual monomer content >0.5 μg/L (USEPA 1980).

VI. Analytical Methods

No standardized method exists for the determination of acrylamide in drinking water. The analytical procedure for the determination of acrylamide (Brown and Rhead 1979) consists of bromination, extraction of the brominated product from water with ethyl acetate, and quantification using high-performance liquid chromatography (HPLC) with an ultraviolet detector. The concentration of the ethyl acetate to dryness and dissolution in a small volume of distilled water before HPLC analysis allows the detection at concentrations of 0.2 μg/L.

VII. Treatment Technologies

Croll et al. (1974) conducted laboratory experiments to determine the effectiveness of conventional treatments, such as coagulation and rapid gravity sand filtration, for removal of acrylamide. Several 400-ml samples of Thames

River water (pH 7.5) containing 25 mg/L kaolin were coagulated by adding 32 mg/L alum and 2 mg/L of an acrylamide-based polymer with a residual acrylamide monomer content of 0.19%. Only about 5% of the residual monomer was removed, suggesting that full-scale water treatment plants using conventional techniques would not be successful in removing acrylamide from drinking water.

Its removal by adsorption was studied by Brown et al. (1980a). Various adsorbants, including granular-activated carbon (GAC) and synthetic resins, were used. GAC may be an effective treatment process, having removed 94 to 96% of the acrylamide from a sample containing 0.5 mg/L and 68 to 70% from a sample containing 10 mg/L. The adsorption of acrylamide was not affected significantly by changes in pH. No significant adsorption was achieved by any of the resins tested, including the XAD-2® resin.

In laboratory experiments by Croll et al. (1974), water containing 6 μg/L acrylamide (at pH 5.0) was treated with 8 mg/L powdered-activated carbon (PAC) and mixed for 30 min. Only 13% of the acrylamide was removed. These data indicated that PAC may not be effective for removing acrylamide from drinking water under the usual conditions in water treatment plants.

No data were found on removal by aeration. Since its Henry's Law constant is 4.38×10^{-3} atm (at 20°C), aeration probably would not be effective.

Croll et al. (1974) evaluated the effects of chemical oxidative treatments on acrylamide removal. Potassium permanganate and ozone were found highly effective. Additional data are needed to optimize these processes. Oxidative degradation products also should be identified and evaluated for toxicity and reactivity.

The selection of individual or combined technologies to achieve acrylamide reduction must be based on a case-by-case technical evaluation and an assessment of the economics involved.

References

Agrawal AK, Seth PK, Squibb RE, Tilson HA, Uphouse LL, Bondy SC (1981) Neurotransmitter receptors in brain regions of acrylamide-treated rats. I: Effects of a single exposure to acrylamide. Pharmacol Biochem Behav 14:527–531.

Agrawal AK, Squibb RE (1981) Effects of acrylamide given during gestation on dopamine receptor binding in rat pups. Toxicol Lett 7:233–238.

Auld RB, Bedwell SF (1967) Peripheral neuropathy with sympathetic overactivity from industrial contact with acrylamide. Can Med Assoc J 96:652–654.

Brown L, Rhead M (1979) Liquid chromatographic determination of acrylamide monomer in natural and polluted aqueous environment. Analyst 104:391–399.

Brown L, Bancroft KCC, Rhead MM (1980a) Laboratory studies on the adsorption of acrylamide monomer by sludge, sediments, clays, peat, and synthetic resins. Water Res 14.779–781.

Brown L, Rhead MM, Bancroft KCC, Allen N (1980b) Model studies of the degradation of acrylamide monomer. Water Res 14·775–779.

Bull RJ, Robinson M, Laurie RD, Stoner GD, Greisiger E, Meier JR, Stober J (1984a) Carcinogenic effects of acrylamide in SENCAR and A/J mice. Cancer Res 44:107–111.

Bull RJ, Robinson M, Stober J (1984b) Carcinogenic activity of acrylamide in the skin and lung of Swiss-ICR mice. Cancer Lett 24:209–212.

Burek JD, Albee RR, Beyer JE, Bell TJ, Carreon RM, Morden DC, Wade CE, Hermann EA, Gorzinski SJ (1980) Subchronic toxicity of acrylamide administered to rats in the drinking water followed by up to 144 days of recovery. J Environ Pathol Toxicol 4:157–182.

Croll BT, Arkell GM, Hodge RPJ (1974) Residues of acrylamide in water. Water Res 8:989–993.

Davenport JG, Farrell DF, Sumi SM (1976) Giant axonal neuropathy caused by industrial chemicals. Neurology 26:919–923.

Dixit R, Seth PK, Mukhtar H (1982) Metabolism of acrylamide into urinary mercapturic acid and cysteine conjugates in rats. Drug Metab Disp 10:196–197.

Edwards PM (1976) The insensitivity of the developing rat foetus to the toxic effects of acrylamide. Chem-Biol Interact 12:13–18.

Frantz SW, Dryzga MD, Freshoar NL, Watanabe PG (1985) *In vivo/in vitro* determination of cutaneous penetration of residual monomer from polyacrylamides. Toxicologist 5:39. (Abstract)

Fullerton PM (1969) Electrophysiological and histological observations on peripheral nerves in acrylamide poisoning in man J Neurol Neurosurg Psychiat 32: 186–192.

Garland TO, Patterson MWH (1967) Six cases of acrylamide poisoning. Brit Med J 4:134–138.

Gorzinski SJ, Morden DC, Albee RR, Wade CE, Carreon RM, Hermann EA, Beyer JE, Burek JD (1979) Results of palatability (12-day) and tolerance (21-day) studies on acrylamide monomer administered in the drinking water to rats. Unpublished report. Submitted to USEPA Office of Toxic Substances by Dow Chemical Co. Midland MI. September 12.

Hashimoto K, Aldridge WN (1970) Biochemical studies on acrylamide, a neurotoxic agent. Biochem Pharmacol 19:2591–2604.

Hashimoto R, Sakamoto J, Tanii H (1981) Neurotoxicity of acrylamide and related compounds and their effects on male gonads in mice. Arch Toxicol 47:179–189.

Igisu H, Goto I, Kawamura Y, Kato M, Izumi K, Kuroiwa Y (1975) Acrylamide encephaloneuropathy due to well water pollution. J Neurol Neurosurg Psychiat 38:581–584.

Ikeda GJ, Miller E, Sapienza PP, Michel TC, King MT, Turner VA, Blumenthal H, Jackson WE III, Levin S (1983) Distribution of ^{14}C-labeled acrylamide and betaine in foetuses of rats, rabbits, beagle dogs and miniature pigs. Food Chem Toxicol 21:49–58.

Johnson KA, Gorzinski SJ, Brodner KM, Campell RA, Wolf CH, Friedman MA, Mast RW (1986) Chronic toxicity and oncogenicity study on acrylamide incorporated in the drinking water of Fischer 344 rats. Toxicol Appl Pharmacol 85: 154–168

Kesson CM, Baird AW, Lawson DH (1977) Acrylamide poisoning. Postgrad Med J 53:16–17.

Kuperman AS (1958) Effects of acrylamide on the central nervous system of the cat. J Pharmacol Exp Therap 28(123): 180–192.

Leswing RJ, Ribelin WE (1969) Physiologic and pathologic changes in acrylamide neuropathy. Arch Environ Hlth 18: 22–29.

Marlowe C, Clark MJ, Mast RW, Friedman MA, Waddell WJ (1986) The distribution of [^{14}C]-acrylamide in male and female Swiss-Webster mice studied by whole body autoradiography. Toxicol Appl Pharmacol 86: 457–465.

McCollister DD, Oyen F, Rowe VK (1964) Toxicology of acrylamide. Toxicol Appl Pharmacol 6: 172–181.

Merigan WH, Barkdoll E, Maurissen JPJ (1982) Acrylamide-induced visual impairment in primates. Toxicol Appl Pharmacol 62: 342–345.

Miller MJ, Carter DE, Sipes IG (1982) Pharmacokinetics of acrylamide in Fischer-344 rats. Toxicol Appl Pharmacol 63: 36–44.

Miller MJ, McQueen CA (1986) The effect of acrylamide on hepatocellular DNA repair. Environ Mutagen 8: 99–108.

Miller MS, Miller MJ, Burks TF, Sipes IG (1983) Altered retrograde axonal transport of nerve growth factor after single and repeated doses of acrylamide in the rat. Toxicol Appl Pharmacol 69: 96–101.

Miller MS, Spencer PS (1984) Single doses of acrylamide reduce retrograde transport velocity. J Neurochem 43: 1401–1408.

NIOSH (1976) National Institute for Occupational Safety and Health. Criteria for a recommended standard . . occupational exposure to acrylamide. U.S. Dept. of Health, Education and Welfare (DHEW), Public Health Service (PHS), Centers for Disease Control (CDC). NIOSH, Washington, DC.

Pryor GT, Uyeno ET, Tilson HA, Mitchell CL (1983) Assessment of chemicals using a battery of neurobehavioral tests: a comparative study. Neurobehavioral Toxicol Teratol 5: 91–117.

Ramsey JC, Young JD, Gorzinski SJ (1984) Acrylamide: toxicodynamics in rats. Unpublished report. Dow Chemical Co., Midland, MI.

Shiraishi Y (1978) Chromosome aberrations induced by monomeric acrylamide in bone marrow and germ cells of mice. Mutat Res 57: 313–324.

Smith MK, Zenick H, Preston R, George EL, Long RE (1986) Dominant lethal effects of subchronic acrylamide administration in the male Long-Evans rat. Mutat Res 173: 273–278.

Thomann P, Koella WP, Krinke G, Petermann H, Zak F, Hess R (1974) The assessment of peripheral neurotoxicity in dogs: comparative studies with acrylamide and clioquinol. Agts Act 4: 47–53.

USEPA[a] (1980) Coagulant aids in the treatment of drinking water. USEPA Office of Drinking Water (ODW), Washington, DC.

USEPA[a] (1984) Miscellaneous synthetic organic chemicals. Occurrence in drinking water, food, and air. USEPA Office of Drinking Water (ODW), Washington, DC.

[a] USEPA documents can be obtained by writing to the appropriate office within the U S Environmental Protection Agency, 401 M Street, S W , Washington, DC 20460 They may also be available by contacting the National Technical Information Service, 5285 Port Royal Road, Springfield, VA 22161 (Telephone 1-800-336-4700)

USEPA[a] (1985a) Draft health effects criteria document for acrylamide. USEPA Criteria and Standards Division (CSD), Office of Drinking Water (ODW), Washington, DC.

USEPA[a] (1985b) Draft technologies and costs for the removal of synthetic organic chemicals from potable water supplies. USEPA Science and Technology Branch, Criteria and Standards Division (CSD), Office of Drinking Water (ODW), Washington, DC.

USEPA[a] (1985c) Assessment of carcinogenic risk of exposure to acrylamide (draft). A staff paper prepared by W.C. Pepelko and J. Cogliano for USEPA Office of Health and Environmental Assessment, Office of Research and Development, Washington, DC.

USEPA (1986) Guidelines for carcinogen risk assessment. Fed Regis 51(185):33992–34003. September 24.

USITC (1984) United States International Trade Commission. Synthetic organic chemicals, U.S. production and sales—1983 U.S. ITC publication no 1588, Washington, DC. pp 255, 268.

Verschueren K (1983) Handbook of environmental data on organic chemicals 2nd Ed. Van Nostrand Reinhold Co., New York, NY.

Windholz M, Budavari S, Blumetti RF, Otterbein ES (eds) (1983) The Merck index—an encyclopedia of chemicals and drugs. 10th Ed. Merck & Co Inc, Rahway, NJ.

Barium[a]

I. General Information

Barium salts are used for a number of purposes, including drilling mud (Kirkpatrick 1978), as a pigment (Miner 1969), and as x-ray contrast medium (Miner 1969). Other uses are summarized by Pidgeon (1964). The properties of barium and its salts are unique to the specific compound. For the structures and properties of barium, barium chloride, and barium sulfate (also known as barite), see Tables 1, 2, and 3, respectively.

Barium is a reactive metal that exists in nature chiefly as the mineral barite ($BaSO_4$) and in much smaller amounts as witherite ($BaCO_3$). These mineral forms are relatively insoluble in water, with high melting and boiling points and very low vapor pressures (Preisman 1964). Barium compounds occur in most geologic materials at levels of 300 to 500 ppm. Barium occurs at low levels in most surface and ground waters at reported levels of less than 340 μg/L. While barium compounds are used commercially in a number of processes, contamination of drinking water is usually the result of naturally occurring barium and not of industrial releases (USEPA 1987).

There are limited survey data on barium in drinking water in the U.S. Most supplies contain less than 200 μg/L. Currently, 60 ground-water supplies and 1 surface water supply exceed the interim Maximum Contamination Level (MCL) of 1,000 μg/L. Barium also occurs in most foods as a low-level contaminant. Based on the limited information available on barium exposure, food seems to be the major source of exposure (USEPA 1987).

II. Pharmacokinetics

A. Absorption

In laboratory animals, the absorption of barium depends upon a number of factors, including species and age (USEPA 1985; Taylor et al. 1962), compound tested (McCauley and Washington 1983), and composition of the diet (Lengemann 1959).

No definitive human barium absorption studies have been reported (USEPA 1985). Barium absorption has been estimated to be approximately 5% in the

[a] Based on USEPA (1985), except for Section I which is based on USEPA (1987)

Table 1. General information and properties of barium

Property	Value
Chemical Abstracts Service (CAS) number	7440-39-3
Chemical formula	Ba
Molecular weight	137.33
Physical state	Silver-white solid
Boiling point	1637–1638°C
Melting point	729–730°C
Density	3.6 g/cm^3 (20°C)
Vapor pressure	1,810 × 10^{-5} mm Hg
Specific gravity	3.6 g/cm^3
Water solubility	Reacts
Log octanol water/partition coefficient	—
Odor threshold	—
Taste threshold	—

Source Chilton (1973), Kirkpatrick (1978), Miner (1969), Pidgeon (1964), Preisman (1964), Reeves (1979)

Table 2 General information and properties of barium chloride[a]

Property	Value
Chemical Abstracts Service (CAS) number	10361-37-2
Chemical formula	$BaCl_2$
Molecular weight	208.24
Physical state	White solid
Boiling point	1,560°C
Melting point	960°C
Density	3.856 g/cm^3 (20°C)
Vapor pressure	—
Specific gravity	3.8 g/km^3
Water solubility	37.5 g/100 mL (26°C)[a]
Log octanol/water partition coefficient	—
Odor threshold	—
Taste threshold	—

Source· Chilton (1973), Kirkpatrick (1978), Miner (1969), Pidgeon (1964); Preisman (1964), Reeves (1979)

[a]Clayton (1981)

Table 3. General information and properties of barium sulfate

Property	Value
Chemical Abstracts Service (CAS) number	7727-43-7
Chemical formula	$BaSO_4$
Molecular weight	233.40
Physical state	Colorless solid
Boiling point	—
Melting point	1,580°C
Density	4.50 g/cm^3 at 20°C
Vapor pressure	—
Specific gravity	4.5 g/cm^3
Water solubility	2.46 mg/L (25°C)[a]
Log octanol/water partition coefficient	—
Odor threshold	—
Taste threshold	—

Source Chilton (1973), Kirkpatrick (1978), Miner (1969), Pidgeon (1964), Preisman (1964), Reeves (1979)

[a]Clayton (1981)

adult (ICRP 1973); however, other data (Harrison et al. 1967) suggest that barium absorption probably is higher. While data in laboratory animals (Lengemann 1959) suggest that barium absorption in children may be significantly greater than in adults, current data are insufficient to resolve the issue.

B. Distribution

In the mouse, intravenously injected barium ($^{133}BaCl_2$) is widely distributed, but is localized principally in the bone (Dencker et al. 1976). Autopsy data indicate that barium levels in human bone are relatively constant and do not appear to increase with age, ranging from an average value of 7.0 ppm in bone at age 0 to 3 mon to 8.5 ppm at age 33 to 74 yr (Sowden and Stitch 1957).

C. Metabolism

The skeletal metabolism of barium in humans is qualitatively similar to that of calcium, although the incorporation of these two elements is quantitatively very different (Bauer et al. 1956, 1957).

D. Excretion

In humans, approximately 72% of ingested barium is eliminated via fecal excretion following oral exposure (Tipton et al. 1966).

III. Health Effects

A. Humans

Acute barium toxicity is associated with hypokalemia and electrocardiographic changes, and other symptoms (Diengott et al. 1964; Gould et al. 1973; Talwar and Sharma 1979). The National Academy of Sciences (NAS 1977) has concluded that the fatal dose of barium chloride for man is about 0.8 to 0.9 g (550 to 600 mg of barium).

In an epidemiology study by Brenniman et al. (1981; see also Brenniman and Levy 1984), the blood pressure of 1,175 males and females, who were exposed to a mean concentration of 7.3 mg/L of drinking water, was compared to the blood pressure of 1,203 males and females who consumed water containing 0.1 mg barium/L. No significant differences in blood pressure ($p < 0.05$) were observed between those who drank water containing 7.3 or 0.1 mg/L. The subjects of the study ranged in age from 18 to more than 75 yr and, in most cases, had lived in their respective communities for a decade or more.

In a human study initiated by USEPA (Wones 1988), 11 healthy men were enrolled in a 10-wk dose–response protocol in which diet was controlled, other aspects of lifestyle known to affect cardiac risk factors were controlled, and the barium content of the drinking water (1.5 L/d) was varied from 0 (first 2 wk) to 5 mg/L (next 4 wk) to 10 mg/L (last 4 wk). Multiple blood and urine samples, morning and evening blood pressures, and 48-hr electrocardiographic monitoring were performed for each dose of barium. There were no changes in morning or evening systolic or diastolic blood pressures, plasma cholesterol, lipoprotein or apolipoprotein levels, serum potassium, total calcium or glucose levels, or urine metanephrine levels. There were no arrhythmias related to barium exposure detected on continuous electrocardiographic monitoring. In summary, drinking water containing barium at 5 and 10 mg/L did not appear to affect any of the cardiovascular risk factors studied.

B. Animals

1. Short-Term Exposure. The acute oral LD_{50} of barium varies markedly with species, compound, age, and other factors (USEPA 1985). For example, the acute oral LD_{50} of barium chloride is 220 mg/kg in weanlings and 132 mg/kg in adult rats (Tardiff et al. 1980).

2. Long-Term Exposure. Schroeder and Mitchener (1975a,b) exposed male and female Long–Evans rats (75 to 80 d old) and Charles River CD strain Swiss mice (19 to 20 d old) to 0 and 5 mg/L barium in drinking water for a lifetime (approximately 0.25 mg/kg/d for rats and 0.825 mg/kg/d for mice). Fifty-two rats/sex/dose were used; and 54 mice/sex/dose were used. No compound-related adverse effects were observed.

Tardiff et al. (1980) exposed 4-wk-old male and female Charles River rats (from 5 to 10/sex/dose) to barium at 0, 10, 50, or 250 ppm (1/3 LD_{50}) in drinking water for 4, 8, and 13 wk (approximately 0, 2.75, 13.7, and 66.25 mg/kg/d at the beginning of the study and 0, 1.7, 6.6, and 31.5 mg/kg/d at the end). Although the barium body burden increased with increasing barium dosage, no compound-related adverse effects (e.g., tissue weights, gross and histopathology and a number of other parameters) were observed in these animals. Blood pressure was not measured.

Perry et al. (1983) exposed female weanling rats to barium at 0, 1, 10, or 100 ppm in drinking water for up to 16 mon (average daily barium doses of 0.051, 0.51, and 5.1 mg/kg, respectively). The control group contained at least 13 animals. With the exception of an increase in blood pressure, there was no evidence of barium-induced adverse health effects at any barium dose level. Systolic blood pressure measurements revealed no increase in pressure in animals exposed to 1 ppm for 16 mon, an increase of 4 mm Hg ($p < 0.01$) in animals exposed to 10 ppm barium for 16 mon, and an increase in systolic pressure of 16 mm Hg ($p < 0.001$) in animals exposed to 100 ppm barium for 16 mon. The animals in this study were maintained in a special contaminant-free environment and fed a rye-based diet (1.5 ppm barium) designed to reduce exposure to trace metals. It is possible that the restricted intake of certain beneficial metals (e.g., calcium) may have predisposed the test animals to the hypertensive effects of barium.

3. Reproductive and Developmental Effects. No adequate mammalian study on the potential reproductive or developmental effects of barium has been identified (USEPA 1985).

4. Mutagenicity and Carcinogenicity. No adequate studies on the mutagenicity or carcinogenicity of barium have been identified (USEPA 1985).

IV. Quantification of Toxicological Effects[a]

A. One-Day Health Advisory

Available data are insufficient to develop a One-Day HA for barium. It is recommended that the Lifetime HA of 5 mg/L (calculated below) be used for all age groups as the One-Day HA.

B. Ten-Day Health Advisory

The available data are insufficient to develop a Ten-Day HA for barium. It is recommended that the Lifetime HA of 5 mg/L (calculated below) be used as the Ten-Day HA for all age groups.

---

[a] See Introduction for discussion of methodology

C. Longer-Term Health Advisory

The available data are insufficient to develop Longer-Term HAs for barium. It is recommended that the Lifetime HA of 5 mg/L be used as the Longer-Term HA for all age groups.

D. Lifetime Health Advisory

USEPA had originally considered using the Perry et al. (1983) rat drinking water study as the basis of the Lifetime HA. However, the Perry et al. (1983) rat study contradicts the results of the Brenniman et al. (1981; see also Brenniman and Levy, 1984) human study. Specifically, barium was without effect on human blood pressure at levels as high as 7.3 mg/L (the highest level tested) in the Brenniman et al. study whereas barium significantly ($p < 0.01$) increased the blood pressure of rats at levels as low as 10 mg/L in the Perry et al. study.

In order to resolve the contradiction between the Perry and Brenniman data, USEPA initiated the Wones (1988) human study. In the Wones human study, barium was without effect upon blood pressure at levels as high as 10 mg/L, supporting the observations of Brenniman et al.

USEPA reasoned that, while the available data would possibly support a Lifetime HA as high as 10 mg/L (i.e., the highest level tested in the Wones study) it was prudent to allow for some level of uncertainty in the data base.

USEPA would normally use an uncertainty factor (UF) of 10 when an HA is derived from human data to allow for the most sensitive population. However, in this case, USEPA believes that a smaller UF is appropriate. Specifically, in relation to hypertension, USEPA believes that adults are the most sensitive population. The data of both Brenniman et al. and Wones were obtained collectively from hundreds of adults who, in some cases, had been exposed to elevated levels of barium in drinking water for decades without adverse effect. Thus, USEPA reasoned that a UF of 10 was not needed to allow for the most sensitive population since that same population was in USEPA's judgment likely included in the Brenniman et al. (1981) study. Based on the weight of all the available data, USEPA believes that a UF of 2 is appropriate in this case. When the Lifetime HA is based on an experimental animal study, USEPA calculates the RfD which, in turn, is used to derive the DWEL, the basis of the Lifetime HA. In this case, however, the basis of the Lifetime HA is human drinking water data.

Based on the previous discussion, USEPA has derived the Lifetime HA from the 10 mg/L level in the Wones human study using a UF of 2. At a drinking water level of 5 mg/L (the Lifetime HA), USEPA believes that appreciably all of daily exposure to barium (e.g., diet, water, etc.) would be due to the ingestion of water.

Determination of the Lifetime Health Advisory

$$\text{Lifetime HA} = \frac{10 \text{ mg/L}}{2} = 5 \text{ mg/L } (5{,}000\ \mu\text{g/L})$$

E. Evaluation of Carcinogenic Potential

No information was located in the available literature regarding the carcinogenic potential of barium in humans nor were any appropriate animal studies found. Using the criteria described in USEPA's guidelines for assessment of carcinogenic risk (USEPA 1986), barium is classified in Group D: not classified. This category is for agents with inadequate evidence of carcinogenicity in humans and animals. The International Agency for Research on Cancer has not evaluated the carcinogenic potential of barium.

V. Other Criteria, Guidance, and Standards

The National Interim Primary Drinking Water Regulations of 1975 established a Maximum Contaminant Level (MCL) drinking water standard of 1 mg/L (USEPA 1976).

The National Academy of Sciences (NAS 1982) derived a 1-d Suggested-No-Adverse-Response Level (SNARL) of 6.0 mg/L. It also derived a chronic SNARL for barium of 4.7 mg/L.

The American Conference of Governmental Industrial Hygienists established an occupational Threshold Limit Value (TLV) of 0.5 mg/m^3 for barium nitrate in air (AGGIH 1980).

The USSR standard for waterborne barium is 4 mg/L (NAS 1977).

The Occupational Safety and Health Administration (OSHA) 8-hr Time-Weighted Average exposure limit for soluble barium compounds is 0.5 mg/m^3 in workplace air (OSHA 1985).

VI. Analytical Methods

Barium is determined by atomic absorption (AA) using either direct aspiration into a flame (USEPA 1979a) or a furnace technique (USEPA 1979b). The direct aspiration AA procedure is a physical method based on barium's absorption of radiation at 553.6 nm. The sample is aspirated into an air-acetylene flame and atomized. A light beam is directed through the flame into a monochromator and onto a detector that measures the amount of light absorbed. Absorbance is proportional to the concentration of barium in the sample, and the detection limit is 100 μg/L.

The furnace AA procedure is similar to direct aspiration except a furnace rather than a flame is used to atomize the sample. The detection limit is 2 μg/L.

VII. Treatment Technologies

Ion exchange, lime softening, and reverse osmosis are effective in removing barium from drinking water. Conventional coagulation/filtration processes are not effective (USEPA 1977).

Weinberg (1973) reported that ion-exchange softening systems are highly efficient (93% to 98%) in reducing barium levels in water, even after water hardness breakthrough. Field data from two Midwestern full-scale ion-exchange softening plants showed that barium removal was comparable to hardness removal on well water containing 11 to 19 mg/L of barium and 225 to 230 mg/L of hardness as $CaCO_3$ (BIF 1970). When these softening units were performing efficiently and removing all of the hardness from the water, they also removed all of the barium.

Lime softening is very effective in removing barium from drinking water. Lime softening achieved >90% removal in the 10 to 11 pH range on well water containing 7 to 8.5 mg/L of naturally occurring barium. Removals decreased below and above this range. Pilot-plant studies conducted at the USEPA Municipal Environmental Research Laboratory and full-scale treatment information on similar types of ground water verified the laboratory data. Pilot-plant test runs on water containing 10 to 12 mg/L of barium at pH 9.2, 10.5, and 11.6 resulted in removals of 84, 93, and 82%, respectively. Grab samples from two full-scale lime softening plants showed removals of 88 and 95%. These plants operated at pH 10.5 and 10.3; the raw water barium concentrations were measured at 7.5 and 17.4 mg/L, respectively (BIF 1970).

A number of studies indicate that reverse osmosis membranes can remove more than 90% of the barium from drinking water. In an experimental long-term study, 99% barium removal was obtained using cellulose acetate membrane at 400 to 800 psi operating pressures (BIF 1970). Other studies by Sorg and Logsdon (1980) achieved similar results, where 95 to 99% removals were obtained by passing water containing 7 mg/L barium through cellulose acetate membranes at 165 to 180 psi operating pressures.

References

ACGIH (1980) American Conference of Governmental Industrial Hygienists Threshold limit values for chemical substances and physical agents in the workroom environment with intended changes for 1980. ACGIH, Cincinnati, OH, p 35.

Bauer GCH, Carlsson A, Lindquist B (1957) Metabolism of ^{140}Ba in man Acta Orth Scand 26:241–254.

Bauer GCH, Carlsson A, Lindquist B (1956) A comparative study of the metabolism of ^{140}Ba and ^{45}Ca in rats. Biochem J 63:535–542.

BIF (1970) Chemicals used on treatment of water and waste water engineering data

BIF, Unit of General Signal Corp, Providence, RI, May.

Brenniman GR, Kojola WH, Levy PS, Carnow BW, Namekata T (1981) High barium levels in public drinking water and its association with elevated blood pressure. Arch Environ Hlth 36(1):28–32.

Brenniman GR, Levy PS (1984) Epidemiological study of barium in Illinois drinking water supplies. In: Calabrese EJ, Tuthill RW, Lyman Condie (eds) Proceedings of conference on inorganics in drinking water and cardiovascular disease. Amherst, MA, May 1984.

Clayton GD, Clayton FE (eds) (1981) Patty's Industrial Hygiene and Toxicology. 3rd Ed. John Wiley and Sons, New York, NY, p 1533.

Dencker L, Nilsson A, Ronnback C, Walinder G (1976) Uptake and retention of ^{133}Ba and ^{140}Ba-^{140}La in mouse tissue. Acta Radiol 15(4):273–287.

Diengott D, Rozsa O, Levy N, Muammar S (1964) Hypokalemia in barium poisoning. Lancet 2:343–344

Gould DB, Sorrell MR, Luperiello AD (1973) Barium sulfide poisoning. Arch Intern Med 132:891–894.

Harrison GE, Carr TEF, Sutton A (1967) Distribution of radioactive calcium, strontium, barium and radium following intravenous injection into a healthy man. Int J Radiat Biol 13(3):235–247.

ICRP (1973) International Commission on Radiological Protection. Alkaline earth metabolism in adult man ICRP publication 20 Hlth Phys 24:125–221.

Kirkpatrick T (1978) Barium compounds. In: Kirk-Othmer encyclopedia of chemical technology, 3rd Ed, Vol 3 John Wiley and Sons, New York, NY. pp 463–479.

Lengemann FW (1959) The site of action of lactose in the enhancement of calcium utilization. J Nutr 69.23–27

McCauley PT and Washington IS (1983) Barium bioavailability as the chloride, sulfate or carbonate salt in the rat. Drug Chem Toxicol 6(2):209–217.

Miner S (1969) Air pollution aspects of barium and its compounds. Technical report. Contract no. PH-22-68-25 Litton Systems, Inc, Bethesda, MD. 69 pp.

NAS (1977) National Academy of Sciences. Drinking water and health, Vol 1. National Academy Press, Washington, DC, pp 207–305.

NAS (1982) National Academy of Sciences. Drinking water and health, Vol 4 National Academy Press, Washington, DC, pp 167–170.

OSHA (1985) Occupational Safety and Health Administration. Code of federal regulations Title 29—Labor. Part 1910—Occupational safety and health standards Subpart Z—Toxic and hazardous substances. Section 1910.1000—Air contaminants. U.S. Government Printing Office, Washington, DC.

Perry HM, Kopp SJ, Erlanger MW, Perry EF (1983) Cardiovascular effects of chronic barium ingestion. In: Hemphill DD (ed). Trace substances in environmental health-XVII. Proceedings of University of Missouri's 17th annual conference on trace substances in environment health University of Missouri Press, Columbia, MO, pp 155–164.

Pidgeon LM (1964) Barium. In. Kirk-Othmer encyclopedia of chemical technology, 2nd ed, Vol 3. John Wiley and Sons, New York, NY, pp 77–80

Preisman L (1964) Barium compounds. In. Kirk-Othmer encyclopedia of chemical technology, 2nd Ed, Vol 3. John Wiley & Sons, New York, NY, pp 80–98

Reeves AL (1979) Barium. In: Friberg L, Nordberg GF, Vouk VB (eds) Handbook on the toxicology of metals. Elsevier/North Holland Biomedical Press, Amsterdam, pp 321–328.

Schroeder HA, Mitchener M (1975a) Life-term effects of mercury, methyl mercury and nine other trace metals on mice. J Nutr 105:452–458.

Schroeder HA, Mitchener M (1975b) Life-term studies in rats. effects of aluminum, barium, beryllium and tungsten. J Nutr 105:421–427.

Sorg TJ, Logsdon GS (1980) Treatment technology to meet the interim primary drinking water regulations for inorganics: part 5. J Am Waterworks Assoc 72(7) 411–422.

Sowden EM, Stitch SR (1957) Trace elements in human tissue. 2. Estimation of the concentrations of stable strontium and barium in human bone Biochem J 67. 104–109.

Talwar KK, Sharma BK (1979) Myocardial damage due to barium chloride poisoning. Indian Heart J 31(4):244–245.

Tardiff RG, Robinson M, Ulmer NS (1980) Subchronic oral toxicity of $BaCl_2$ in rats J Environ Path Tox 4:267–275.

Taylor DM, Bligh PH, Duggan MH (1962) The absorption of calcium, strontium, barium and radium fron the gastrointestinal tract of the rat. Biochem J 83:25–29.

Tipton IH, Stewart PL, Martin PG (1966) Trace elements in diets and excreta. Hlth Phys 12:1683–1689.

USEPA[a] (1976) National interim primary drinking water regulations. Report no EPA-570/9-76-003. USEPA Office of Drinking Water, Washington DC.

USEPA[a] (1977) Manual of treatment techniques for meeting the interim primary drinking water regulations, revised. Report no. EPA-600/8-77-005. USEPA Office of Drinking Water, Washington, DC.

USEPA[a] (1979a) Method 208.1. Atomic absorption, direct aspiration. In: Methods for chemical analysis of water and wastes. Report no. EPA-600/4-79-020. March. USEPA Environmental Monitoring and Support Laboratory, Cincinnati, OH

USEPA[a] (1979b) Method 208.2. Atomic absorption, furnace technique. In: Methods for chemical analysis of water and wastes Report no. EPA-600/4-79-020. March. USEPA Environmental Monitoring and Support Laboratory, Cincinnati, OH.

USEPA[a] (1985) Draft health effects criteria document for barium. USEPA Criteria and Standards Division, Office of Drinking Water, Washington, DC.

USEPA (1986) Guidelines for carcinogenic risk assessment. Fed Regis 51(185): 33992–34003. September 24.

USEPA[a] (1987) Estimated national occurrence and exposure to barium in public drinking water supplies. USEPA Criteria and Standards Division, Office of Drinking Water, Washington, DC.

Weinberg LM (1973) Report of analytical evaluation and treatability study For Wight

[a]USEPA documents can be obtained by writing to the appropriate office within the U S Environmental Protection Agency, 401 M Street, S W , Washington, DC 20460 They may also be available by contacting the National Technical Information Service, 5285 Port Royal Road, Springfield, VA 22161 (Telephone. 1-800-336-4700)

Consulting Engineers on Lake Zurich Water Well #5. CHEMED Corp., Dearborn Environmental Engineers, Dearborn, MI. July.

Windholz M (ed) (1976) The Merck index—an encyclopedia of chemicals and drugs. 9th Ed. Merck & Co Inc, Rahway, NJ, p 995.

Wones J (1988) Lack of effect of drinking water barium on cardiovascular risk factors. Unpublished study. University of Cincinnati, College of Medicine, Cincinnati, OH.

Cadmium[a]

I. General Information

Cadmium is used for a number of purposes, including batteries, electroplating, in pigments, and as an alloy with other metals (Stubbs 1978). The properties of some common cadmium compounds are given in Tables 1–3.

Cadmium is a naturally occurring metallic element, present in most of the earth's crust at levels below 1 ppm. It is obtained commercially as a by-product in zinc production. It is released to the environment during commercial activities; however, these releases have not resulted in the contamination of ground and surface waters (USEPA 1987).

Naturally occurring levels of cadmium in surface and ground waters are in the range of 1 to 10 μg/L. State monitoring data indicate that 21 ground-water supplies and 4 surface-water supplies currently exceed 10 μg/L. Cadmium occurs at low levels in food and air. The Food and Drug Administration Total Diet Study reports that an American adult ingests 34 μg/d in the diet. Thus, food appears to be the major route of exposure for cadmium (USEPA 1987).

Cadmium is found in both cigarettes and cigarette smoke. Because the absorption of inhaled cadmium can approach levels as high as 96% (CEC 1978), smoking accounts for a substantial fraction of the body burden of cadmium in cigarette smokers (Ellis et al. 1979).

II. Pharmacokinetics

A. Absorption

The absorption of cadmium following oral administration to laboratory animals, and presumably humans, is modified by many factors, including dose (Engstrom and Nordberg 1979), age (Kostial et al. 1983), diet (Suzuki et al. 1969), and the presence of other metals such as calcium (Washko and Cousins 1976). Cadmium does not readily penetrate the skin (CEC 1978), but it is very readily absorbed following inhalation. As much as 96% of the cadmium deposited in the lungs may be absorbed (CEC 1978).

[a] Based on USEPA (1985).

Reviews of Environmental Contamination and Toxicology, Vol. 107

Table 1. General information and properties of cadmium

Property	Value
Chemical Abstracts Service (CAS) number	7440-43-9
Chemical formula	Cd
Molecular weight	112.40
Physical state	Soft white solid
Boiling point	765°C
Melting point	320.9°C
Density	8.642 g/cm^3
Vapor pressure	1.4 mm Hg (400°C)
Specific gravity	8.6 g/cm^3
Water solubility	Insoluble in water
Log octanol/water partition coefficient	—
Odor threshold	—
Taste threshold	—

Source: IARC (1976), Parker (1978), Schindler (1967), Weast (1971)

Table 2. General information and properties of cadmium chloride

Property	Value
Chemical Abstracts Service (CAS) number	10108-64-2
Chemical formula	$CdCl_2$
Molecular weight	183.32
Physical state	Solid
Boiling point	—
Melting point	568°C
Density	4.047 g/cm^3
Vapor pressure	—
Specific gravity	4.04 g/cm^3
Water solubility	1.4 kg/L (20°C)[a]
Log octanol/water partition coefficient	—
Odor threshold	—
Taste threshold	—

Source IARC (1976), Parker (1978), Schindler (1967), Weast (1971)
[a]Clayton (1981).

Table 3. General information and properties of cadmium oxide

Property	Value
Chemical Abstracts Service (CAS) number	1306-19-0
Chemical formula	CdO
Molecular weight	128.40
Physical state	Solid
Boiling point	1,559°C
Melting point	900°C
Density	8.15 g/cm^3
Vapor pressure	—
Specific gravity	8.15 g/cm^3
Water solubility	5.0 mg/L (25°C)[a]
Log octanol/water partition coefficient	—
Odor threshold	—
Taste threshold	—

Source: IARC (1976); Parker (1978), Schindler (1967), Weast (1971).
[a] Seidell (1940).

B. Distribution

In both rats (Sabbioni et al. 1978) and humans (Sumunio et al. 1975), cadmium distributes throughout the body and accumulates in the kidney and liver, where it may attain levels 10 to 100 times greater than those of other tissues.

C. Metabolism

Unlike the typical organic drinking water contaminants, whole cadmium is not metabolized to other compounds. Once within the body, cadmium readily combines with the low-molecular-weight protein metallothionein (Foulkes 1982).

D. Excretion

Cadmium is eliminated in humans principally via the urine (USEPA 1985). Elimination of cadmium is slow in humans probably because of its marked accumulation and retention in the kidney and liver (Friberg et al. 1974). The half-life for elimination of cadmium has been estimated to be 10 to 33 yr (Ellis et al. 1979). In humans, average body retention of radiolabeled cadmium chloride, measured 1 to 5 wk after exposure, was approximately 4.6% (McLellan et al. 1978).

III. Health Effects

A. Humans

In humans, the symptoms of cadmium toxicity following acute exposure include nausea, vomiting, diarrhea, muscular cramps, and salivation (Arena 1963). In the case of severe intoxication, sensory disturbances, liver injury, and convulsions may result. In fatal intoxication, these are followed by shock and/or renal failure and cardiopulmonary depression (CEC 1978). The estimated acute lethal dose of cadmium is 350 to 35,000 mg for a 70-kg adult (CEC 1978). The No-Observed-Adverse-Effect Level (NOAEL) for emesis in adults is 0.043 mg/kg/d following acute oral exposure to cadmium salts (Lauwerys 1979).

Chronic nonoccupational oral exposure to very high levels of cadmium has resulted in Itai-Itai disease observed in Japan (principally in multiparous women), which is characterized by pain, osteomalacia, osteoporosis, proteinuria, glucosuria, and anemia (USEPA 1985).

Cadmium may play a role in hypertension, but there is considerable debate concerning what role that may be (Kopp et al. 1982, 1983; Perry et al. 1977a,b).

Renal toxicity (e.g., proteinuria) following low-level chronic oral exposure is believed to be the most sensitive manifestation of cadmium toxicity (CEC 1978; USEPA 1985). The concentration at which 10% of the population is likely to display signs of renal dysfunction has been estimated to be 180 to 220 μg Cd/g renal cortex. Individuals with tissue concentrations over 285 μg/g usually display signs of renal dysfunction (USEPA 1985). Friberg et al. (1974) hypothesized that renal damage may occur when, over a 50-yr period, a person's daily cadmium intake equals or exceeds 0.352 mg/d.

B. Animals

1. Short-Term Exposure. The acute oral LD_{50} of cadmium compounds in the rat ranges from 16 mg/kg for cadmium cyanide to >5,000 mg/kg for cadmium sulfide (CEC 1978). Toxic effects of oral exposure to various cadmium compounds have been observed in a variety of animal tissues (USEPA 1985) including the nervous system (Gabbiani et al. 1967), kidney (CEC 1978), liver (Stowe et al. 1972), bone (Larsson and Piscator 1971), hematopoietic system (Stowe et al. 1972), cardiovascular system (Kopp et al. 1978), and immune system (Koller 1973).

2. Long-Term Exposure. Cadmium-induced renal toxicity (e.g., proteinuria) has been observed in animals in the absence of renal histopathology (CEC 1978). In a 24-wk male rat drinking water study, animals exposed to 2.15 and

6.44 mg Cd/kg/d developed proteinuria ($p < 0.05$), while animals exposed to the lowest level tested, 0.84 mg Cd/kg/d (NOAEL), did not (Kotsonis and Klaassen 1978). In a 12-mon rat drinking water study, no adverse effects were observed in animals exposed to 0.008, 0.035, 0.181, 0.361, or 0.375 mg/kg/d (No-Observed-Effect Level, or NOEL). However, at 3 mon, the animals exposed to the highest level tested, 3.04 mg Cd/kg/d, developed anemia and did not gain weight normally (Decker et al. 1958).

3. Reproductive Effects. In a rat oral study, cadmium was administered at 0, 0.1, 1.0, and 10.0 mg/kg/d (as $CdCl_2$) to groups of male and female adult rats for 6 wk. Males and females were mated for 3 wk, and cadmium was administered during the mating period. Pregnant females were also given cadmium during the gestation period. In the 10-mg/kg group, the number of total implants and live fetuses decreased significantly ($p < 0.05$), while resorptions increased significantly ($p < 0.01$). Fetal body weight was significantly decreased ($p < 0.05$) and fetuses showed delayed ossification of the sternebrae and caudal vertebrae. No effects were observed at 0.1 or 1.0 mg/kg/d (Sutou et al. 1980).

4. Developmental Effects. In addition to the fetal effects described above by Sutou et al. (1980), Ahokus et al. (1980) observed fetal growth retardation when rat dams were exposed to 100 mg/L but not 0.1 or 10 mg/L during gestation (Ahokas et al. 1980).

5. Mutagenicity. While cadmium has been observed to cause chromosomal aberrations in several in vitro studies (e.g., DiPaolo and Casto 1979; Watanabe et al. 1979), strong evidence of mutagenic effects following oral ingestion is not available (USEPA 1985).

6. Carcinogenicity. Cadmium and cadmium compounds have been shown to induce sarcomas at local injection sites (Gunn et al. 1967; Haddow et al. 1964). In addition, cadmium chloride administered to rats by aerosol for 18 mon produced lung tumors (Takenaka et al. 1983). These data may not be relevant to oral exposure to cadmium in drinking water (USEPA 1985).

Although cancers of the prostate and lung have been noted in cadmium smelter workers in an epidemiological study (Lemen et al. 1976), evidence regarding the carcinogenicity of cadmium in humans following oral exposure is largely conjectural (USEPA 1985).

No evidence of cadmium oncogenicity has been found in chronic oral animal studies (Kanisawa and Schroeder 1969; Loser 1980; Schroeder et al. 1965).

IV. Quantification of Toxicological Effects[a]

A. One-Day Health Advisory

The study by Lauwerys (1979) was selected to serve as the basis for calculating the One-Day HA for cadmium. In this study, the NOAEL for cadmium-induced emesis in adult humans following a single dose was 0.043 mg/kg/d. This study was selected because it is of appropriate duration and was conducted in humans; more suitable data are not available.

The One-Day HA for a 10-kg child is calculated as follows:

$$\text{One-Day HA} = \frac{(0.043 \text{ mg/kg/d})(10 \text{ kg})}{(10)(1 \text{ L/d})} = 0.043 \text{ mg/L } (43\ \mu\text{g/L})$$

B. Ten-Day Health Advisory

The 24-wk oral exposure study in rats by Kotsonis and Klaassen (1978) was considered for use as the basis of the Ten-Day HA. In this study, a NOAEL of 0.84 mg/kg/d was identified, based on the absence of proteinuria. If this NOAEL and an uncertainty factor of 100 were used, the Ten-Day HA value would be 0.08 mg/L. This value is not markedly different from the One-Day HA of 0.043 mg/L (based on a study that demonstrated cadmium-induced emesis in adult humans). However, since the Ten-Day HA value of 0.08 mg/L would be greater than the One-Day HA value, it is recommended that the more conservative One-Day HA of 0.043 mg/L (43 μg/L) be used as the Ten-Day HA.

C. Longer-Term Health Advisory

The available data are insufficient to develop Longer-Term HAs for cadmium. It is recommended that the Drinking Water Equivalent Level (DWEL) of 18 μg/L be used as the Longer-Term HA for a 70-kg adult and the modified DWEL of 5 μg/L be used as the Longer-Term HA for a 10-kg child.

D. Lifetime Health Advisory

There are no adequate oral exposure studies in humans that provide a NOAEL for the chronic effects of cadmium. Friberg et al. (1974) concluded that the critical concentration of cadmium in the renal cortex of humans associated with renal dysfunction is 200 μg/g wet weight; this is supported by the recent reassessment by Kjellstrom et al. (1984). The 200 μg/g critical concentration was based on a comprehensive review of evidence from animal experiments and from analyses of kidneys from workers occupationally exposed to cadmium. The 200 μg/g value is probably the most widely accepted estimate of

[a] See Introduction for a discussion of methodology.

the critical concentration for renal dysfunction (NAS 1977; CEC 1978). However, Roels et al. (1983) reported that the critical concentration in the human renal cortex is 216 μg/g tissue wet weight and that <10% of occupationally exposed males may develop renal dysfunction at this concentration.

Several models have been proposed to estimate the daily intake of (exposure to) cadmium required to produce the critical concentration in the renal cortex. Each model has inherent limitations. Friberg et al. (1974) estimated that a daily cadmium intake of 0.352 mg/d for 50 yr would result in a renal cortex concentration of 200 μg/g. This model assumes 4.5% absorption of the daily oral dose and 0.01% excretion/d of the total body burden. Thus, 0.352 mg/d in a 70-kg adult (0.005 mg/kg/d) is a reasonable estimate of the daily cadmium intake that would result in renal dysfunction. Since 0.005 mg/kg/d is associated with renal dysfunction, it is considered to be a Lowest-Observed-Adverse-Effect Level (LOAEL) value. The use of a LOAEL in calculation of a Lifetime HA normally requires that an uncertainty factor of 100 be used. However, given the relatively low level of uncertainty concerning cadmium toxicity in this case, an uncertainty factor of 100 is unreasonably high. A factor of 10 is more appropriate for calculation of the Lifetime HA.

Using a LOAEL of 0.005 mg/kg/d, the Lifetime Health Advisory is derived as follows:

Step 1: Determination of the Reference Dose (RfD)

$$\text{RfD} = \frac{(0.005\ \text{mg/kg/d})}{(10)} = 0.0005\ \text{mg/kg/d}$$

The uncertainty factor of 10, although smaller than normally required by NAS/ODW guidelines, was judged to best reflect the uncertainty concerning cadmium toxicity in humans.

Step 2: Determination of the Drinking Water Equivalent Level (DWEL)

$$\text{DWEL} = \frac{(0.0005\ \text{mg/kg/d})(70\ \text{kg})}{(2\ \text{L/d})} = 0.018\ \text{mg/L}\ (18\ \mu\text{g/L})$$

Step 3: Determination of the Lifetime Health Advisory

$$\text{Lifetime HA} = (0.018\ \text{mg/L})\ (25\%) = 0.005\ \text{mg/L}\ (5\ \mu\text{g/L})$$

E. Evaluation of Carcinogenic Potential

A quantitative evaluation of the carcinogenicity of cadmium following oral exposure has not been made since there is no conclusive evidence that cadmium is carcinogenic following oral exposure. USEPA has recommended that cadmium not be considered a suspect human carcinogen for the purpose of calculating an ambient water quality criterion (USEPA 1980). Based on exposure to cadmium via inhalation, the International Agency for Research on Cancer (IARC 1982) has classified cadmium and certain cadmium compounds in Group 2B: limited evidence of carcinogenicity in humans and sufficient evidence of carcinogenicity in animals.

Using the criteria described in USEPA's guidelines for assessment of carcinogenic risk (USEPA 1986), cadmium on the basis of inhalation data may be classified in Group B1: probable human carcinogen. This category is for agents for which there is limited evidence of carcinogenicity from human studies and sufficient evidence from animal studies. However, since data are inadequate to conclude that cadmium is carcinogenic via ingestion, cadmium in drinking water may be assigned to Group D: not classified. This category is for agents with inadequate animal evidence of carcinogenicity.

V. Other Criteria, Guidance, and Standards

The National Academy of Sciences (NAS 1982) has calculated a 1-d Suggested-No-Adverse-Response Level (SNARL) of 0.150 mg/L for cadmium in drinking water for 70-kg adults. It has also calculated a 7-d SNARL of 0.021 mg/L and a chronic exposure SNARL of 0.005 mg/L of drinking water for a 70-kg adult (NAS 1982). The latter value is based on the assumption that water contributes 20% of the daily cadmium intake. A Food Additives Organization/World Health Organization expert committee has proposed a provisional tolerable weekly standard of no more than 57.1 to 71.4 μg/wk (WHO 1972). WHO has also recommended that the concentration of cadmium in drinking water not exceed 0.005 mg/L. The Commission of the European Communities (CEC 1975) concurs with that recommendation. The current USEPA primary drinking water standard for cadmium is 0.010 mg/L of drinking water (USEPA 1976). The recommended Threshold Limit Value (TLV) for cadmium dusts, salts, and oxide fumes for occupational 8-hr time-weighted average exposure is 0.05 mg/m^3 (ACGIH 1980). The Occupational Safety and Health Administration (OSHA) 8-hr Time-weighted Average (TWA) exposure limit for cadmium fume is 0.1 mg/m^3 in workplace air with an acceptable ceiling concentration of 0.3 mg/m^3 (OSHA 1985). The OSHA 8-hr TWA exposure limit for cadmium dust is 0.2 mg/m^3 in workplace air with an acceptable ceiling concentration of 0.6 mg/m^3 (OSHA 1985).

VI. Analytical Methods

Cadmium levels are determined by atomic absorption (AA) using either direct aspiration into a flame (USEPA 1979a) or a furnace technique (USEPA 1979b). The direct-aspiration AA procedure is a physical method based on cadmium's absorption of radiation at 228.8 nm. The sample is aspirated into an air-acetylene flame and atomized. A light beam is directed through the flame into a monochromator and onto a detector that measures the amount of light absorbed. Absorbance is proportional to the concentration of cadmium in the sample. The detection limit is 5 μg/L. The furnace AA procedure is similar to direct aspiration, except a furnace rather than a flame is used to atomize the sample. The detection limit is 0.1 μg/L.

VII. Treatment Technologies

Effective removal of cadmium from source waters may be achieved with treatment methods such as coagulation with alum or iron salts, lime softening, ion exchange, and reverse osmosis. Laboratory experiments and pilot-plant studies indicate that the effectiveness of cadmium removal by coagulation is pH-dependent. Ferric sulfate coagulation studies on river water containing 0.3 mg/L of cadmium showed removal efficiencies increased from 20% at pH 7.2 to above 90% at pH 8 and above. Alum coagulation results on river water also increased with increased pH level, but the data indicated that above pH 8, removals may depend on the turbidity of the raw water. In some tests with low-turbidity water (1 to 10 tu), removals decreased as the pH increased (USEPA 1978).

Lime softening is capable of achieving greater than 98% cadmium removal from water in the pH range of well water containing 0.3 mg/L of cadmium. Comparable removal efficiency was achieved at pH 11.2 to 11.3 when the initial cadmium concentration was increased up to 10 mg/L (USEPA 1978).

There are limited performance data on the use of ion exchange as a treatment method for removal of cadmium from drinking water. The plating industry uses ion exchange for reducing cadmium in wastewaters, and other wastewater streams studied have successfully used ion exchange for removing cadmium (Amax 1977; Ameron 1978; Laszlo 1977; Lindstedt et al. 1971; Nippon 1976). There is one report of 99% removal efficiency for cadmium from drinking water using a home ion-exchange softener (Culligan Co. 1982). Tap water spiked with 0.10 mg/L of cadmium chloride and used as feed water to a cation exchanger on the sodium cycle produced water with a cadmium level less than 0.01 mg/L.

Reverse osmosis can effectively remove cadmium from drinking water. A study by Mixon (1973) showed a 90% and 98% cadmium removal, respectively, from 0.10 mg/L and 0.98 mg/L spiked water samples, using three laboratory-scale cellulose acetate membranes operated at 400 psi. No differences in cadmium rejection were noted when barium, chromium, copper, lead, and zinc were introduced. A study performed by Huxstep (1982) in Florida related to inorganic contaminant removal from potable water by reverse osmosis resulted in a 96% to 98% removal of cadmium.

Protection against cadmium from corrosion of water distribution systems may be achieved by a number of methods, including pH adjustment, addition of lime, increasing alkalinity, or addition of phosphates or silicates. The extent and type of treatment selected is dependent on the characteristics of the water and the compatibility of existing treatment with various materials used to convey the water through the distribution system.

References

ACGIH (1980) American Conference of Governmental Industrial Hygienists. Documentation of the threshold limit values, 4th Ed. ACGIH, Cincinnati, OH. pp 59–61.

Ahokas RA, Dilts PV, LaHaye EB (1980) Cadmium-induced fetal growth retardation: protective effect of excess dietary zinc. Am J Obstet Gynecol 136:216–226.

Amax Inc (1977) Removal of metal ions from wastewater. U.S. Patent 4,025,430, submitted January 12, 1976. May 24.

Ameron Inc (1978) System for removal of toxic heavy metals from drinking water. U.S. Patent 4,096,064, submitted April 5, 1976 June 20.

Arena JM (1963) Poisoning: chemistry, symptoms, and treatment. Charles C Thomas, Springfield, IL, p 127.

CEC (1975) Commission of the European Communities. Proposal for a council directive relating to the quality of water for human consumption. J Offic Eur Commun 18.2–17.

CEC (1978) Commission of the European Communities. Criteria (dose/effect relationships) for cadmium Pergamon Press, Oxford, pp 1–198.

Clayton GD, Clayton FE (eds) (1981) Patty's Industrial Hygiene and Toxicology. 3rd Ed. John Wiley and Sons, New York NY, p 1565.

Culligan Company (1982) Personal communication to V.J. Ciccone Engineering. Aug 1, 1982.

Decker CF, Byerrum RU, Hoppert CA (1957) A study of the distribution and retention of cadmium-115 in the albino rat. Arch Biochem Biophys 66:140–145.

DiPaola JA, Casto BC (1979) Quantitative studies of *in vitro* morphological transformation of Syrian hamster cells by inorganic metal salts. Cancer Res 39:1008–1013.

Ellis KJ, Vartsky D, Zanzi I, Cohn SH, Yasumura S (1979) Cadmium. *in vivo* measurement in smokers and nonsmokers. Science 205:323–325.

Engstrom B, Nordberg GF (1979) Dose dependence of gastrointestinal absorption and biological half-time of cadmium in mice. Toxicology 13·215–222.

Foulkes EC (ed) (1982) Biological roles of metallothionein. Elsevier/North Holland Press, New York, NY

Friberg L, Piscator M, Nordberg GF, Kjellstrom T (1974) Cadmium in the environment, 2nd Ed. CRC Press Inc, Boca Raton, FL.

Gabbiani G, Gregory A, Baic D (1967). Cadmium-induced selective lesions of sensory ganglia. J Neuropathol Exp Neurol 26:498–506.

Gunn SA, Gould TC, Anderson WAD (1967) Specific response of mesenchymal tissue to carcinogenesis by cadmium. Arch Pathol 83:493–499.

Haddow A, Roe FJC, Dukes CE, Mitchley BCV (1964) Cadmium neoplasia: sarcomata at the site of injection of cadmium sulphate in rats and mice. Br J Cancer 18:667–673.

Huxstep MR (1982) Inorganic contaminant removal from potable water by reverse osmosis (Task 49AS, Treatment of small community water supplies by reverse osmosis). Charlotte Harbor (Florida) Water Association, Inc., Progress Report, January 1–March 31, 1982. U.S. Environmental Protection Agency.

IARC (1976) International Agency for Research on Cancer. Monographs on the evaluation of carcinogenic risk of chemicals to man. Cadmium, nickel, some epoxides, miscellaneous industrial chemicals and general considerations on volatile anesthetics, Vol 11. IARC, Lyon, France, pp 39–74.

IARC (1982) International Agency for Research on Cancer. IARC monographs on the evaluation of the carcinogenic risk of chemicals to humans. Suppl 4:133–135. IARC, Lyon, France.

Kanisawa M, Schroeder HA (1969) Life term studies on the effect of trace elements on spontaneous tumors in mice and rats. Cancer Res 29:892–895.

Kjellstrom T, Elinder CG, Friberg L (1984) Conceptual problems in establishing the critical concentration of cadmium in human kidney cortex. Environ Res 33:284–295.

Koller LD (1973) Immumosuppression produced by lead, cadmium, and mercury. Am J Vet Res 34:1457–1458.

Kopp SJ, Fisher VW, Erlanger M, Perry EF, Perry HM (1978) Electrocardiographical, biochemical and morphological effects of chronic low level cadmium feeding on rat heart. Proc Soc Exp Biol Med 159:339–345.

Kopp SJ, Glonek T, Perry HM, Erlanger M, Perry EF (1982) Cardiovascular actions of cadmium at environmental exposure levels. Science 217:837–839.

Kopp SJ, Perry HM, Perry EF, Erlanger M (1983) Cardiac physiologic and tissue metabolic changes following chronic low-level cadmium and cadmium plus lead ingestion in the rat. Toxicol Appl Pharmacol 69:149–160.

Kostial K, Simonovic I, Rabar I, Blanusa M, Landeka M (1983) Age and intestinal retention of mercury and cadmium in rats. Environ Res 31:111–115.

Kotsonis FN, Klaassen CD (1978) The relationship of metallothionein to the toxicity of cadmium after prolonged oral administration to rats. Toxicol Appl Pharmacol 46:39–54.

Larsson SE, Piscator M (1971) Effect of cadmium on skeletal tissue in normal and cadmium-deficient rats. Isr J Med Sci 7:495–498.

Laszlo M (1977) Process for removing heavy metals from fluid media. U.S. Patent 4,060,410, submitted July 7, 1975. November 29.

Lauwerys R (1979) Cadmium in man. In: Webb (ed) The chemistry, biochemistry, and biology of cadmium. Elsevier/North Holland Biomedical Press, Amsterdam, pp 433–453.

Lemen RA, Lee JS, Wagoner JK, Blejer HP (1976) Cancer mortality among cadmium production workers. Ann NY Acad Sci 271:273–279.

Lindstedt KD, Houck CP et al. (1971) Trace element removals in advanced wastewater treatment processes. Water Pollut Cont Fed 43(7):1507–1513.

Loser E (1980) A 2-year oral carcinogenicity study with cadmium on rats. Cancer Lett 9:191–198.

McLellan JS, Flanagan PR, Chamberlain MJ, Valberg LS (1978) Measurement of dietary cadmium absorption in humans. J Toxicol Environ Hlth 4.131–138.

Mixon FO (1973) Removal of toxic metals from water by reverse osmosis. R&D

Progress report no. 889 U S. Dept Int, Off of Saline Water, Washington, DC.

NAS (1977) National Academy of Sciences. Drinking water and health, Vol 1 National Academy Press, Washington DC, p 939.

NAS (1982) National Academy of Sciences. Safe Drinking Water Committee. Drinking water and health, Vol 4. National Academy Press, Washington, DC pp 170–174.

Nippon Electric Co, Ltd (1976) Improvements in or relating to the extraction of heavy metals from industrial wastewaters. British Patent 1,457,528, submitted December 19, 1972. December 1, 1976

OSHA (1985) Occupational Safety and Health Administration. Code of federal regulations. Title 29—Labor Part 1910—Occupational safety and health standards. Subpart Z—Toxic and hazardous substances. Section 1910 1000—Air contaminants U.S. Government Printing Office, Washington, DC

Parker PD (1978) Cadmium compounds. In: Kirk-Othmer encyclopedia of chemical technology. 3rd Ed, Vol 4 John Wiley & Sons, New York, pp 387–411

Perry HM, Erlanger M, Perry EF (1977a) Elevated systolic pressure following chronic low-level cadmium feeding. Am J Physiol 232: H114–H121.

Perry HM, Erlanger M, Perry EF (1977b) Hypertension following chronic, very low dose cadmium feeding. Proc Soc Exp Biol Med 156: 173–176.

Ribelin WE (1963) Atrophy of rat testis as index of chemical toxicity Arch Pathol 75. 229–235.

Roels R, Lauwerys R, Dardenne AN (1983) The critical level of cadmium in human renal cortex: a re-evaluation Toxicol Lett 15: 357–360.

Sabbioni E, Marafante E, Amantini L, Ubertalli L, Pietra R (1978) Cadmium toxicity studies under long term-low level exposure (LLE) conditions. I Metabolic patterns in rats exposed to present environmental dietary levels of Cd for two years. Sci Total Environ 10: 135–161.

Schindler PW (1967) Heterogenous equilibria involving oxides, hydroxides, carbonates and hydroxide carbonates. In: Equilibrium concepts in natural water systems Adv in Chem Series 67. American Chemical Society, Washington, DC, pp 196–221

Schroeder HA, Balassa JJ, Vinton WH (1965) Chromium, cadmium, and lead in rats. effects on life span, tumors and tissue levels. J Nutr 86: 51–66.

Seidell A (1940) Solubilities of inorganic and metal organic compounds 3rd Ed D. Van Nostrand Co., New York, NY, pp 444–445.

Stowe HD, Wilson M, Goyer RA (1972) Clinical and morphologic effects of oral cadmium toxicity in rabbits Arch Pathol 94 · 389–405.

Stubbs RL (1978) Cadmium—the metal of benign neglect. Proceedings of the 1st international cadmium conference. Metal Bulletin Ltd, London, England, pp 7–12.

Sumino K, Hayakawa K, Shibata T, Kitamura S (1975) Heavy metals in normal Japanese tissues Arch Environ Hlth 30: 487–494.

Sutou S, Yamamoto K, Sendota H, Sugiyama M (1980) Toxicity, fertility, teratogenicity and dominant lethal tests in rats administered cadmium subchronically. III. Fertility, teratogenicity, and dominant lethal test. Exotoxicol Environ Saf 4: 51–56

Suzuki S, Taguchi T, Yokohashi G (1969) Dietary factors influencing upon the retention rate of orally administered $^{115}Cd\ Cl_2$ in mice with special reference to calcium and protein concentrations in diet Ind Hlth 7: 155–162.

Takenaka S, Oldiges H, Konig H, Hochrainer O, Oberdorster G (1983) Carcinogenicity of cadmium chloride aerosols in W rats. J Nat Cancer Inst 70:367–373.

USEPA[a] (1976) National interim primary drinking water regulations. USEPA Office of Water Supply, Washington, DC, pp 59–62.

USEPA[a] (1978) Manual of treatment techniques for meeting the interim primary drinking water regulations, revised. Report no. EPA-600/8-77-005. USEPA Office of Drinking Water, Washington, DC.

USEPA[a] (1979a) Water Method 213.1. Atomic absorption, direct aspiration. Report no. EPA-600/4-79-020. In: Methods for chemical analysis of water and wastes. March. USEPA Environmental Monitoring and Support Laboratory, Cincinnati, OH.

USEPA[a] (1979b) Method 213.2. Atomic absorption, furnace technique. Report no. EPA-600/4-79-020. In: Methods for chemical analysis of water and wastes. March. USEPA Environmental Monitoring and Support Laboratory, Cincinnati, OH.

USEPA[a] (1980) Ambient water quality criteria for cadmium. Report no. EPA-440/5-80-025. USEPA Office of Water Regulation and Standards. Washington, DC.

USEPA[a] (1985) Final draft of the drinking water criteria document on cadmium. USEPA Office of Drinking Water (ODW), Washington, DC.

USEPA[a] (1986) Guidelines for carcinogen risk assessment. Fed Regis 51(185):33992–34003. September 24.

USEPA[a] (1987) Occurrence of cadmium in public water supplies. USEPA Criteria and Standards Division (CSD), Office of Drinking Water, Washington, DC.

Washko PW, Cousins RJ (1976) Metabolism of ^{109}Cd in rats fed normal and low-calcium diets. J Toxicol Environ Hlth 1:1055–1066.

Watanabe T, Shimada T, Endo A (1979) Mutagenic effects of cadmium on mammalian oocyte chromosomes. Mutat Res 67:349–356.

Weast RC (ed) (1971) CRC handbook of chemistry and physics, 52nd Ed. The Chemical Rubber Co, Cleveland, OH.

WHO (1972) World Health Organization. Evaluation of certain food additives and the contaminants mercury, lead, and cadmium. Sixteenth Report of the Joint FAO/WHO Expert Committee on Food Additives. WHO Technical Report Series no. 505, Food and Agriculture Organization (FAO) Nutrition Meetings Report Series no. 51. WHO, Geneva, Switzerland.

WHO (1984) World Health Organization. Guidelines for drinking water quality—recommendations; Vol 1. WHO, Geneva, Switzerland.

[a] USEPA documents can be obtained by writing to the appropriate office within the U.S. Environmental Protection Agency, 401 M Street, S W , Washington, DC 20460 They may also be available by contacting the National Technical Information Service, 5285 Port Royal Road, Springfield, VA 22161 (Telephone 1-800-336-4700)

Chromium[a]

I. General Information

Chromium and its salts have a variety of uses. Hexavalent chromium compounds are used widely in industry for chrome alloy and chromium metal production, for metal finishing and corrosion control (Love 1947), and as mordants in the textile industry (Iler 1954). Chromium salts are used as anticorrosive agents in cooling waters, in the leather tanning industry, in the manufacture of catalysts, in pigments and paints, and in fungicides and wood preservatives (Hartford 1979). The properties of chromium (Cr), chromium (III) chloride, and chromic (VI) acid are given in Tables 1–3.

Chromium is a relatively rare, naturally occurring element in the earth's crust. It occurs in most rocks and minerals at levels of 200 ppm. A few minerals contain chromium at levels of 2,000 to 3,000 ppm. Chromium is not mined commercially in the U.S., but it is imported. Chromium is released to the environment during industrial activities. However, current data suggest that surface and ground water levels of chromium are the result of naturally occurring chromium leaching from mineral deposits. Soluble chromium has been reported in surface waters at levels up to 84 $\mu g/L$ and in ground water at levels of 50 $\mu g/L$ (USEPA 1987).

Federal surveys of surface and ground-water drinking water supplies have reported that most supplies contain less than 5 $\mu g/L$. Currently, 17 ground-water supplies and 1 surface-water supply are known to exceed USEPA's interim standard of 50 $\mu g/L$ (USEPA 1987).

II. Pharmacokinetics

A. Absorption

With the exception of Cr III glucose tolerance factor (GTF), an organic complex of Cr III, Cr VI is more readily absorbed than Cr III. In humans and experimental animals, gastrointestinal absorption of inorganic salts of Cr III is low (from 0.5 to 3%). However, Cr VI and organic complexes of Cr III are

[a] Based on USEPA (1985).

Table 1. General information and properties of chromium

Property	Value
Chemical Abstracts Service (CAS) number	7440-47-3
Chemical formula	Cr
Molecular weight	51.996
Physical state	Blue-white solid
Boiling point	2,642°C
Melting point	1,900°C
Density	7.14 g/cm^3
Vapor pressure	—
Specific gravity	7.14 g/cm^3
Water solubility	0.5 μg/L
Log octanol/water partition coefficient	—
Odor threshold	—
Taste threshold	—

Source. Hem (1970); Weast (1971); Windholz (1976).

Table 2. General information and properties of chromium (III) chloride

Property	Value
Chemical Abstracts Service (CAS) number	10025-73-7
Chemical formula	$CrCl_3$
Molecular weight	122.90
Physical state	Solid
Boiling point	—
Melting point	83°C
Density	2.76 g/cm^3 (15°C)
Vapor pressure	—
Specific gravity	2.76 g/cm^3 (15°C)
Water solubility	58.36–73.26 g/100 mL sat sol (25°C)[a]
Log octanol/water partition coefficient	—
Odor threshold	—
Taste threshold	—

Source: Hem (1970); Weast (1971); Windholz (1976)
[a]Seidell (1940).

Table 3. General information and properties of chromic (VI) acid, dipotassium salt

Property	Value
Chemical Abstracts Service (CAS) number	7789-00-6
Chemical formula	K_2CrO_4
Molecular weight	194.20
Physical state	Solid
Boiling point	—
Melting point	968.3°C
Density	2.732 g/cm³ (18°C)
Vapor pressure	—
Specific gravity	2.7 g/cm³ (18°C)
Water solubility	62.9 g/100 mL (20°C)
Log octanol/water partition coefficient	—
Odor threshold	—
Taste threshold	—

Source: Hem (1970); Weast (1971); Windholz (1976).

more readily absorbed (approximately 2 to 10% for Cr VI and 10 to 25% for organic complexes of Cr III) (USEPA 1985).

In humans administered 20 ng of Cr III as $^{51}CrCl_3$ in water, 0.5% of the administered dose was recovered in the urine. Since urinary recovery can be assumed to be proportional to absorption, it appears very little absorption in humans occurs (Donaldson and Berreras 1966). In rats, Mertz et al. (1965) reported 2 to 3% absorption of Cr III based on a total body count of animals administered $^{51}CrCl_3$ by intubation at doses ranging from 1.5 to 100 μg/kg.

GTF was absorbed in rats at 10 to 25% of the administered dose (Mertz 1976; Mertz et al. 1978). An estimate of 2.1% absorption of Cr VI based on recovery in urine was reported for humans administered 20 ng of $Na_2^{51}CrO_4$ in water (Donaldson and Berreras 1966). Tissue levels of chromium were 12.5 times higher in rats administered drinking water containing 25 mg/L Cr III as chromic chloride than in rats that drank water containing 25 mg/L Cr VI as potassium chromate (MacKenzie et al. 1958).

B. Distribution

Compounds containing chromium III and VI differ in their patterns of distribution of Cr III/Cr VI within an organism. In general, Cr III crosses membranes much more slowly than Cr VI (USEPA 1985). Chromium circulates in the plasma primarily in a nondiffusible form. A small fraction (9 to 12%) is in a more diffusible form, which is filtered and partially reabsorbed in the kidney (Collins et al. 1961). An approximate plasma half-life of 6 hr for ^{51}Cr III in rats was reported by Hopkins (1965) after intravenous administration of either 0.1 or 1.0 μg/kg. Cr III has an affinity for iron-binding proteins (Gray and Sterling 1950; Hopkins and Schwarz 1964).

The spleen and kidneys were shown to have the highest concentrations of chromium when rats were administered Cr III as chromium chloride in intravenous doses of 0.1 or 1.0 μg/kg (Hopkins 1965). Similar results were reported by MacKenzie et al. (1958) when rats drank water containing 25 mg/L of either Cr III (as chromic chloride) or Cr VI (as potassium chromate). The calculated doses were 1.87 mg/kg/d for males and 2.41 mg/kg/d for females.

The placenta appears to be highly selective in its permeability to the various forms of chromium. Inorganic Cr III administered as $^{51}CrCl_3$ intravenously or by stomach intubation does not cross the placental barrier to an appreciable extent in rats (Mertz et al. 1969). However, Cr III administered by stomach intubation to pregnant rats in the form of GTF, obtained from yeast, is recovered readily fron the fetus (Mertz and Roginski 1971). The dosages in these two studies were unspecified.

Cr VI traverses biological membranes by diffusion or facilitated transport, possibly via an anion transport system (Alexander et al. 1982). It is reduced to Cr III intracellularly by the cytochrome P-450 system in the presence of nicotinamide adenine dinucleotide phosphate (NADPH). Cr III reacts with nucleophilic ligands and cellular macromolecules (Gruber and Jennette 1978).

C. Metabolism

The metabolism of chromium in mammalian species is not well understood. It is complicated by the presence of the two oxidation states, Cr III and Cr VI (USEPA 1985).

D. Excretion

The kidney appears to be the principal route of excretion of chromium compounds. More than 80% of absorbed chromium is excretion in the urine (Krainz and Talmage 1952). Very little is known about the form in which chromium is excreted.

After intravenous administration, chromium is also excreted in the feces, although reports in the literature vary considerably on the percentage. Hopkins (1965) reported that 0.5 to 1.7% of the initial dose of Cr III was excreted in the feces of rats 8 hr after intravenous administration of $^{51}CrCl_3$ at 0.1 μg/100 g.

III. Health Effects

A. Humans

In general, Cr VI compounds are more toxic than Cr III compounds. Cr VI has been shown to produce liver and kidney damage, internal hemorrhage, dermatitis, and respiratory problems. The immediate symptoms are generally nausea, repeated vomiting, and diarrhea (USEPA 1985). Doses of 0.5 to 1.5 g of $K_2Cr_2O_7$ have been fatal in humans. The estimated LD_{LO} for $K_2Cr_2O_7$ in children is 26 mg/kg (Cr VI at 9.2 mg/kg) (NIOSH 1983).

Subchronic and chronic dermal exposure to Cr VI in the form of chromic acid may cause contact dermatitis and ulceration of the skin (Burrows 1978). For example, Denton et al. (1954) reported data on an individual who was patch-tested on three occasions with 0.005% $K_2Cr_2O_7$ solution and the filtrate of two cement samples, which contained 0.0001% and 0.0004% Cr VI. The individual repeatedly showed a positive erythematous, edematous, papulovesicular patch-test reaction to each test solution.

Chronic inhalation of dust or air containing Cr VI may cause respiratory problems, including perforated or ulcerated nasal septa and decreased spirometric values (USEPA 1985). For example, Bloomfield and Blum (1928) reported perforated/ulcerated nasal septa and inflamed nasal mucosa in workers exposed to chromic (VI) acid (0.1 to 5.6 mg/m^3) for 1 wk to 3 yr.

B. Animals

1. Short-Term Exposure. In general, Cr VI compounds are more toxic than Cr III compounds. The oral LD_{50} for various salts of Cr III ranges from 600 to 2,600 mg/kg (Smyth et al. 1969). The oral LD_{50} of Cr VI (as $Na_2Cr_2O_7$) in rats is 19.8 mg/kg (NIOSH 1983).

Gross and Heller (1946) exposed both male and female rats for 60 d to drinking water containing K_2CrO_4 at either 300 or 500 mg/L (Cr VI at 80 and 134 mg/L, respectively). Using reported average body weights of 270 and 260 g, respectively, and assuming consumption of 28 mL water/d, the average ingested doses of Cr VI are calculated to be 8.3 and 14.4 mg/kg/d, respectively. After 2 mon, the rats receiving Cr VI at 8.3 mg/kg/d were described as normal. A "slight roughness of coat" was noted in rats receiving 14.4 mg/kg/d, though this is not considered to be an adverse health effect and is not associated with other adverse health effects.

2. Long-Term Exposure. In a 1-yr rat drinking water study, consumption of water containing 0 to 25 mg/L of either Cr III ($CrCl_3$) or Cr VI (K_2CrO_4) (0 to 1.87 mg/kg/d for males and 0 to 2.41 mg/kg/d for females) produced no significant differences in weight gain, appearance, or pathological changes in the blood or other tissues (MacKenzie et al. 1958). No-Observed-Adverse-Effect Levels (NOAELs) of 1.87 mg/kg/d for males and 2.41 mg/kg/d for females can be identified from the results of this study.

In a rat drinking water study in which 5 mg/L Cr III (about 0.42 mg/kg/d) was administered from the time of weaning until death, no adverse effects were observed (Schroeder et al. 1965). A NOAEL of 0.42 mg/kg/d can be identified from the results of this study.

In a 4-yr female dog drinking water study (5 dose groups with 2 animals per group), Cr VI (K_2CrO_4) at 0.45 to 11.2 mg/L (0.012 to 0.30 mg/kg Cr VI) did affect physical condition, food consumption, growth rate, organ weights, urinalysis results, or hematological analyses. Therefore, a NOAEL of 0.30 mg/kg/d can be identified from the results of this study (Anwar et al. 1961).

3. Reproductive and Development Effects. No information was located on the reproductive and developmental effects of chromium.

4. Mutagenicity. The genotoxic effects of chromium are well documented in both in vivo and in vitro studies. The pathway by which chromium exerts these effects is believed to involve penetration of the cell membrane by Cr VI, followed by intracellular reduction to Cr III. Extracellular Cr III also crosses the cell membrane but less efficiently than Cr VI. Once inside the cell, Cr III can form tight complexes with DNA, an affinity that accounts for its mutagenic potential (USEPA 1985).

Compounds of both Cr III and Cr VI increase noncomplementary nucleotide incorporation into DNA (Raffetto et al. 1977; Majone and Rensi 1979), though Cr VI produces effects at lower doses. Exposure of cells from rat liver and kidney to Cr VI leads to increased cross-linking in DNA. Petrilli and De Flora (1978) reported positive Ames tests for Cr VI. However, Cr III exerted no effect at relatively high concentrations, presumably because of its relatively poor ability to penetrate cells. Similar results were reported by Gentile et al. (1981).

The difficulty in observing mutagenic effects of Cr III may be related to its slight uptake by cells under most conditions. Warren et al. (1981) studied the mutagenicity of a series of hexacoordinate Cr III compounds and concluded that, in the proper ligand environment, the metal possesses considerable genetic toxicity.

5. Carcinogenicity. Inadequate evidence is available to determine whether oral exposure to chromium can lead to cancer. No increase in tumor rates over those of control animals was observed in rats fed Cr III (chromium oxide pigments) at 293, 586, or 1,466 mg/kg/d for 2 yr (Ivankovic and Preussman 1975). And, while the carcinogenicity of inhaled Cr VI is well established for occupational exposure of humans (Hayes et al. 1979), the effects are observed only in the respiratory passages and lungs. The effects on the respiratory tract are believed to have no bearing on carcinogenic risk following oral exposure to chromium (USEPA 1985).

IV. Quantification of Toxicological Effects[a]

During the assessment of chromium compound toxicity, it is important to remember that chromium III is an essential nutrient required in trace quantities for normal glucose metabolism, i.e., GTF. Some forms of chromium may also be important in the metabolism of lipids and other carbohydrates (USEPA 1985).

The following Health Advisories are based on the effects of Cr VI *measured as total chromium*. Separate Health Advisories were not established for Cr III for the following reasons:

1. According to Schroeder and Lee (1975), oxidizing agents (e.g., due to chlorination of water) do not accelerate the normal conversion of Cr III to Cr VI at the point of consumption (i.e., the tap).
2. Health Advisories based on total chromium allow for the possible conversion of Cr III to Cr VI.
3. Cr VI is probably more toxic than Cr III. Thus, Health Advisories based on the effects of Cr VI conservatively protect against the toxic effects of any Cr III that is not converted to Cr VI.

A. One-Day Health Advisory

The available data are insufficient to develop a One-Day HA for total chromium. It is recommended that the Ten-Day HA of 1.4 mg/L be used as the One-Day HA for a 10-kg child.

B. Ten-Day Health Advisory

Based on the NOAEL (14.4 mg/kg/d) for Cr VI from the Gross and Heller (1946) study, the Ten-Day HA for a 10-kg child is calculated as follows:

$$\text{Ten-Day HA} = \frac{(14.4\ \text{mg/kg/d})(10\ \text{kg})}{(100)(1\ \text{L/d})} = 1.4\ \text{mg/L}\ (1400\ \mu\text{g/L})$$

C. Longer-Term Health Advisory

MacKenzie et al. (1958) studied the effects of chronic ingestion of Cr III and Cr VI in male and female Sprague–Dawley rats (34 d old). Given an average weight of 375 g for males and 290 g for females and an average daily water intake of 28 mL (Arrington 1972), the average dose for males and females receiving 25 mg/L is 1.87 and 2.41 mg/kg/d, respectively. The only significant adverse effect noted was an approximate reduction in water consumption in the animals receiving the highest dose (25 mg/L) of Cr VI. With a NOAEL

[a] See Introduction for discussion of methodology

of 2.41 mg/kg/d from this study, the Longer-Term HAs are calculated as follows:

For a 10-kg child:

$$\text{Longer-Term HA} = \frac{(2.41\ \text{mg/kg/d})(10\ \text{kg})}{(100)(1\ \text{L/d})} = 0.24\ \text{mg/L}\ (240\ \mu\text{g/L})$$

For a 70-kg adult:

$$\text{Longer-Term HA} = \frac{(2.41\ \text{mg/kg/d})(70\ \text{kg})}{(100)(2\ \text{L/d})} = 0.84\ \text{mg/L}\ (840\ \mu\text{g/L})$$

D. Lifetime Health Advisory

The MacKenzie et al. study (1958) is also appropriate to use in calculating the Lifetime HA. The Anwar et al. (1961) study was not selected because only two animals were included in each dose group.

With a NOAEL of 2.41 mg/kg/d, the Lifetime HA is derived as follows:

Step 1: Determination of the Reference Dose (RfD)

$$\text{RfD} = \frac{(2.41\ \text{mg/kg/d})}{(100)(5)} = 0.0048\ \text{mg/kg/d}$$

where 100 = uncertainty factor, chosen in accordance with NAS/ODW guidelines for use with a NOAEL from an animal study; 5 = additional uncertainty factor to compensate for less-than-lifetime exposure.

Step 2: Determination of the Drinking Water Equivalent Level (DWEL)

$$\text{DWEL} = \frac{(0.0048\ \text{mg/kg})(70\ \text{kg})}{(2\ \text{L/d})} = 0.17\ \text{mg/L}\ (170\ \mu\text{g/L})$$

Step 3: Determination of Lifetime Health Advisory

$$\text{Lifetime HA} = (170\ \mu\text{g/L})\ (71\%) = 120\ \mu\text{g/L}$$

where 71% = estimated relative source contribution from water.

E. Evaluation of Carcinogenic Potential

No evidence of carcinogenic effects has been noted following oral exposure to chromium. Therefore, no assessment for carcinogenic risks from oral exposure to chromium have been conducted. Inhalation of chromium, however, is associated with an increased frequency of lung cancer in humans.

USEPA's Carcinogen Assessment Group (CAG) has estimated the lifetime cancer risk due to a constant exposure to air containing 1 $\mu g/m^3$ of elemental chromium to be 1.2×10^{-2} (USEPA 1983).

Based on exposure to chromium via inhalation, the International Agency for Research in Cancer (IARC 1982) has classified chromium and certain chromium compounds in Group 1: sufficient evidence for carcinogenicity in humans and animals. Using the criteria described in USEPA's guidelines for assessment of carcinogenic risk (USEPA 1986), chromium is classified in Group A: human carcinogen. This category is for agents for which there is sufficient evidence to support the causal association between exposure to the agents and cancer. However, since there are inadequate data to conclude that chromium is carcinogenic via ingestion, chromium in the context of drinking water may be classified in Group D: not classified. This category is for agents with inadequate animal evidence of carcinogenicity.

V. Other Criteria, Guidance, and Standards

The U.S. Public Health Service has established the level of 50 μg Cr VI/L as a standard for drinking water (USPHS 1962). The National Academy of Sciences has established a similar level of 50 μg total chromium/L for drinking water (NAS 1974; USEPA 1976). The National Institute for Occupational Safety and Health (NIOSH) has established a standard of 1 $\mu g/m^3$ for carcinogenic forms of Cr VI in workplace air (NIOSH 1975). NIOSH has also established a Time-Weighted-Average (TWA) of 25 $\mu g/m^3$ or a 50 mg/m^3 ceiling for noncarcinogenic forms of Cr VI in workplace air. The recommended ambient water quality criterion for Cr VI is 50 μg/L (USEPA 1980). An estimated adequate and safe intake for chromium of 50 to 200 μg/d for adults has been established by NAS (NAS 1980a,b) based on the absence of signs of chromium deficiency in the major portion of the U.S. population, which consumes an average of 60 μg chromium/d. The Occupational Health and Safety Administration (OSHA) has set an 8-hr TWA exposure limit of 0.5 mg/m^3 for chromium, soluble chromic salts, and chromous salt exposure.

VI. Analytical Methods

Chromium levels are determined by atomic absorption (AA) using either direct aspiration into a flame (USEPA 1979a) or a furnace technique (USEPA 1979b). The direct-aspiration AA procedure is a physical method based on chromium's absorption of radiation at 357.9 nm. The sample is aspirated into an air-acetylene flame and atomized. A light beam is directed through the flame into a monochromator, and then onto a detector that measures the amount of light absorbed. Absorbance is proportional to the concentration of chromium in the sample. The detection limit is 50 μg/L.

The furnace AA procedure is similar to direct aspiration except that a furnace rather than a flame is used to atomize the sample. The detection limit is 1 μg/L.

VII. Treatment Technologies

The available treatment technologies for removing chromium from water include coagulation/filtration, lime softening, ion exchange, and reverse osmosis. Which treatment is appropriate depends on the species of chromium present.

In laboratory and pilot-plant studies using ferric sulfate, nearly 100% of Cr III was removed in the range of pH 6.5 to 9.5. Alum was less effective between pH 7.5 and 8.5, with removal efficiency around 90% or better. Above and below this pH range, removal efficiency was slightly lower (80 to 90%). In similar tests analyzing Cr VI removal, only one coagulant, ferrous sulfate, was effective. When alum and ferric sulfate were used, Cr VI removal efficiencies did not exceed 30%. By comparison, ferrous sulfate coagulation achieved 90% removal efficiency or better (USEPA 1977).

Results of jar and pilot plant tests indicate that Cr III removal efficiencies with lime softening was approximately 72% at pH 8.5 to 9.5 and greater than 99% at pH 11 to 11.5. Results with Cr VI in the same tests in all cases indicated less than 10% removal (USEPA 1977; Sorg 1979).

Since Cr III occurs in cationic species and Cr VI in anionic species, a cation exchanger in series with an anion exchanger may be required for their removal. Removal of Cr VI from sewage (Sorg 1979), industrial wastewater, rinse waters from chromium-plating operations (Miller and Mindler 1978), cooling tower blowdown (Richardson et al. 1968; Miller and Mindler 1978), and metal recovery (Sussman et al. 1945) has been demonstrated. Laboratory tests on simulated Arizona well water (total dissolved solids 174 mg/L, pH 7.85) containing 0.019 mg/L of Cr VI showed a breakthrough of Cr VI at roughly 12,000 bed volumes (USEPA 1982). Reports concerning industrial wastewater treatment indicate that ion exchange can successfully remove Cr III to below 0.05 mg/L (Patterson 1975). Strong acid cationic resins have been used for removing Cr III as a contaminant from metal-plating rinse waters and from chromate-treated cooling waters. Vendor information indicates that operating pH levels of between 6 and 8 are adequate for Cr III removal: A pH above 7 is slightly better than a pH below 7 (Rohm and Haas Co 1980). Ion-exchange softening using standard strong acid synthetic resins operating in the sodium cycle should effectively remove Cr III with 90% or greater efficiency (Amore 1977). In tests of home softeners with tap water spiked with 1 mg/L of chromium nitrate, the chromium content continued to be reduced to 0.020 mg/L after 192 cycles, at which point the test was discontinued.

Reverse osmosis (RO) membranes can efficiently remove from 82 to 99% of the chromium in a feedwater source (Mixon 1973; Johnston et al. 1978). Pilot plant tests using both cellulose acetate and hollow fiber (polyamide) membranes demonstrated the effectiveness of the technologies in removing both Cr III and Cr VI. Cr III removal ranged from 90 to 98% and Cr VI removal from 82 to 97%.

References

Alexander J, Aseth J, Norseth T (1982) Uptake of chromium by rat liver mitochondria. Toxicology 24: 115–122.

Amore F (1977) Technical letter 20: removal of water supply contaminants—chromium. Illinois State Water Survey. Illinois Environmental Protection Agency, Springfield.

Anwar RA, Langham RF, Hoppert CA, Alfredson BV, Byerrum RU (1961) Chronic toxicity studies: III. Chronic toxicity of cadmium and chromium in dogs. Arch Environ Hlth 3: 456–460.

Arrington LR (1972) The laboratory animals. In: Introductory laboratory animal science. The breeding, care and management of experimental animals. Interstate Printers and Publishers, Danville, Illinois, pp 9–11.

Burrows D (1978) Chromium and the skin. Br J Dermatol 99: 587–595.

Collins RJ, Fromm PO, Collings WD (1961) Chromium excretion in the dog. Am J Physiol 201: 795–798.

Davids HW, Lieber M (1951) Underground water contamination by chromium wastes. Water Sewage Works 98: 528–534.

Donaldson RM Jr, Barreras RF (1966) Intestinal absorption of trace quantities of chromium. J Lab Clin Med 68: 484–493.

Gentile JM, Hyde K, Schubert J (1981) Chromium genotoxicity as influenced by complexation and rate effects. Toxicol Lett 7: 439–448.

Gray SJ, Sterling K (1950) The tagging of red cells and plasma proteins with radioactive chromium. J Clin Invest 29: 1604–1613.

Gross WG, Heller VG (1946) Chromates in animal nutrition. Ind Hyg Toxicol 28: 52–56.

Gruber JE, Jennett KW (1978) Metabolism of the carcinogen chromate by rat liver microsomes. Biochem Biophys Res Commun 82: 700–706.

Hartford WH (1979) Chromium compounds. In: Grayson M and Eckroth D (eds) Kirk-Othmer encyclopedia of chemical technology, Vol 6. John Wiley and Sons, New York, NY, pp 82–120.

Hayes RB, Lilienfeld AM, Snell LM (1979) Mortality in chromium chemical production workers: a prospective study. Int J Epidemiol 8: 365–374.

Hem JD (1970) Study and interpretation of the chemical characteristics of natural water, 2nd Ed. U.S. Geological Survey Water-Supply Paper 1473. p 199.

Hopkins LL (1965) Distribution in the rat of physiological amounts of injected Cr^{51} (III) with time. Am J Physiol 209: 731–735.

Hopkins LL, Schwarz K (1964) Chromium (III) binding to serum proteins, specifically siderophilin. Biochem Biophys Acta 90: 484–491.

IARC (1982) International Agency for Research on Cancer. IARC monographs on the evaluation of the carcinogenic risk of chemicals to humans. Supplement 4: 133–135. IARC, Lyon, France.

Iler RK (1954) Process for the production of Verner type chromium complexes. U.S. Patent no. 2,683,156.

Ivankovic S, Preussman R (1975) Absence of toxic and carcinogenic effects after administration of high doses of chromic oxide pigment in subacute and long-term feeding experiments in rats. Food Cosmet Toxicol 13: 347–351.

Johnston JK, Lim HS (1978) Removal of persistent contaminants from municipal effluents by reverse osmosis. Environmental Protection Service, Environment Canada.

Kraintz L, Talmage RV (1952) Distribution of radioactivity following intravenous administration of trivalent chromium-51 in the rat and rabbit. Proc Soc Exp Biol Med 81:490–492.

Love CH (1947) German production of some of the more important inorganic pigments. Hobart Publishing Co, Washington, DC. pp 47–63.

MacKenzie RD, Byerrum RU, Decker CF, Hoppert CA, Langham RF (1958) Chronic toxicity studies. II. Hexavalent and trivalent chromium administered in drinking water to rats. AMA Arch Ind Hlth 18:232–234.

Majone F, Rensi D (1979) Mitotic alterations, chromosome aberrations and sister chromatid exchanges induced by hexavalent and trivalent chromium on mammalian cells *in vitro*. Caryologia 32:379–392.

Mertz W, Roginski EE, Reba RC (1965) Biological activity and fate of trace quantities of intravenous chromium (III) in the rat. Am J Physiol 109:489–494.

Mertz W, Roginski EE, Feldman FJ, Thurman DE (1969) Dependence of chromium transfer into the rat embryo on the chemical form. J Nutr 99:363–367.

Mertz W, Roginski EE (1971) Chromium metabolism: the glucose tolerance factor. In: Mertz W and Cornatzer WE (eds) Newer trace elements in nutrition. Marcel Dekker, New York, NY, pp 123–151.

Mertz W (1976) Chromium and its relation to carbohydrate metabolism. In: Burch RE, Sullivan JF (eds) Symposium on trace elements. Med Clin North Am 60.739–744.

Mertz W, Anderson RA, Wolf WR, Roginski EE (1978) Progress in chromium nutrition research. In: Kirchgessner M (ed) Trace element metabolism in man and animals. Proc Third Int Symp. Freising. July 1977, pp 272–278.

Miller WS, Mindler AB (1978) Ion exchange separation of metal ions from water and waste waters. Permutit R&D Center.

Miller WS (1978) Removal and recovery of chromates from cooling tower blowdown. In: Ion Exchange for Pollution Control. Vol I. CRC Press, Inc. Cleveland. OH.

Mixon FO (1973) The removal of toxic metals from water by reverse osmosis. INT-OSWRDPR-73-899. U.S. Department of the Interior, Washington, DC.

NAS (1974) National Academy of Sciences. Water quality criteria, 1972. National Academy Press, Washington, DC, p 62.

NAS (1980a) National Academy of Sciences. Recommended dietary allowances, 9th revised Ed. National Academy Press, Washington, DC, pp 159–161.

NAS (1980b) National Academy of Sciences. Drinking water and health, Vol 3. National Academy Press, Washington, DC, pp 266, 364–369, 374–375.

NIOSH (1975) National Institute for Occupational Safety and Health. Occupational exposure to chromium. VI. Criteria document, U.S. Department of Health, Education and Welfare (USDHEW) NIOSH, Washington, DC, pp 76–129.

NIOSH (1983) National Institute for Occupational Safety and Health. Registry of Toxic Effects of Chemical Substances (RTECS). Vol 2. NIOSH, Washington, DC, p 72.

OSHA (1985) Occupational Safety and Health Administration. Code of Federal Regulations. Title 29—Labor. Part 1910—Occupational safety and health standards. Subpart Z—Toxic and hazardous substances. Section 1910.1000—Air con-

taminants. U.S. Government Printing Office, Washington, DC.
Patterson JW (1975) Wastewater treatment technology. Ann Arbor Science Publisher, Inc, Ann Arbor, MI.
Petrilli FL, De Flora S (1978) Oxidation of inactive trivalent chromium to the mutagenic hexavalent form. Mutat Res 58: 167–173.
Raffetto G, Parodi S, Parodi C, DeFarrari M, Troiano R, Brambilla G (1977) Direct interaction with cellular targets as the mechanism for chromium carcinogenesis. Tumori 63: 503–512.
Richardson EW, Stobbe EC, Bernstein S (1968) Ion exchange traps chromates for reuse. Environ Sci Technol 2(11): 1006–1016.
Rohm and Haas Co (1980) Amberlite Ion Exchange Resins. Technical Bulletins for IR-120 and IRA-402.
Schroeder HA, Balassa JJ, Vinton WH Jr. (1965) Chromium, cadmium and lead in rats: Effects on life span, tumors and tissue levels. J Nutr 86: 51–66.
Schroeder DC, Lee GF (1975) Potential transformations of chromium in natural waters. Water Air Soil Pollut 4: 355–365.
Seidell A (1940) Solubilities of Inorganic and Metal Organic Compounds. 3rd Ed. D. Van Nostrand Co., New York, NY, pp 444–445.
Smyth HF, Carpenter CP, Weil CS, Pozzanni UC, Striegel JA, Nycum JS (1969) Range finding toxicity data: List VII. Am Ind Hyg Assoc J 30: 470.
Sorg TJ (1979) Treatment technology to meet the interim primary drinking water regulations for inorganics: part 4. J Am Waterworks Assoc 71(8): 454–466.
Sussman S, Nachod FC, Wood W (1945) Metal recovery by anion exchange, Ind Eng Chem 37(7): 618–622.
USEPA[a] (1976) National interim primary drinking water regulations. Report no. EPA-570/9-76-003. USEPA Office of Drinking Water, Washington DC, pp 63–64.
USEPA[a] (1977) Manual of treatment techniques for meeting the interim primary drinking water regulations, revised. Report no. EPA-600/8-77-005. USEPA Office of Drinking Water, Washington, DC.
USEPA[a] (1979a) Method 218.1. Atomic absorption, direct aspiration. In: Methods for chemical analysis of water and wastes, Report no. EPA-600/4-79-020. March 1979. USEPA Office of Drinking Water, Washington, DC.
USEPA[a] (1979b) Method 218.2. Atomic absorption, furnace technique. In: Methods for chemical analysis of water and wastes, Report no. EPA-600/4-79-020. March 1979. USEPA Office of Drinking Water, Washington, DC.
USEPA[a] (1980) Ambient water quality criteria for chromium. Report no. EPA-440/5-80-035. USEPA Office of Water, Washington, DC.
USEPA[a] (1982) Personal communication. USEPA Municipal Environmental Research Laboratory (MERL), Cincinnati, OH.
USEPA[a] (1983) Health assessment document for chromium (review draft). Report no. EPA-600/8-82-014A. USEPA Office of Health Effects Assessment, Washington, DC.

[a] USEPA documents can be obtained by writing to the appropriate office within the U.S. Environmental Protection Agency, 401 M Street, S.W., Washington, DC 20460. They may also be available by contacting the National Technical Information Service, 5285 Port Royal Road, Springfield, VA 22161 (Telephone. 1-800-336-4700).

USEPA[a] (1985) Health effects criteria document for chromium. USEPA Criteria and Standards Division (CSD), Office of Drinking Water (ODW). Washington, DC.

USEPA (1986) Guidelines for carcinogen risk assessment. Fed Regis 51(185):33992–34003. September 24.

USEPA (1987) Estimated national occurrence and exposure to chromium in public drinking water supplies. Criteria and Standards Division (CSD), Office of Drinking Water (ODW), Washington, DC.

USPHS (1962) U.S. Public Health Service. Drinking water standards. U.S. Public Health Service publication no. 956. U.S. Government Printing Office, Washington, DC, pp 36–39.

Warren G, Schultz P, Bancroft D, Bennett K, Abbott EH, Rogers S (1981) Mutagenicity of a series of hexacoordinate chromium (III) compounds. Mutat Res 90:111–118.

Weast R (ed) (1971) Handbook of chemistry and physics. 52nd Ed. CRC Press, Cleveland, OH. pp B-65, B-83–84, B-122, B-137.

Windholz M (ed) (1976) The Merck index—an encyclopedia of chemicals and drugs. 9th Ed. Merck & Co Inc, Rahway, NJ, pp 228–289.

Cyanide[a]

I. General Information

Cyanides are a group of organic and inorganic compounds that contain the cyano (CN) group. Free cyanide is defined as the sum of cyanide present as hydrogen cyanide (HCN) and as CN^-. The organic cyanides are called nitriles; a few of them dissociate to yield CN^- or HCN. In this Health Advisory, only a few widely used industrial inorganic cyanides—hydrogen cyanide (HCN, or prussic acid), sodium cyanide (NaCN), and potassium cyanide (KCN)—are discussed.

Cyanide is used in rat and other mammalian pest poisons; silver and metal polishes; photographic solutions; fumigation products; in the production of various resins such as acrylates, methyl acrylate, and nitriles; and in electroplating. The structure and properties of several cyanide compounds are given in Tables 1–3.

In 1978, cyanide production in the U.S. exceeded 700 million lb (317.1 million kg). Cyanide wastes are released from the pyrolysis of natural and synthetic materials (Towill et al. 1978). Despite numerous potential sources of pollution, cyanide is relatively uncommon in U.S. drinking water. In 1970, a survey of 969 water supplies failed to find cyanide concentrations above 0.2 mg/L. Of the 2,595 samples examined, the highest cyanide concentration found was 8 μg/L and the average was 0.09 μg/L (McCabe et al. 1970).

II. Pharmacokinetics

A. Absorption

Free cyanide is readily absorbed from the lungs, gastrointestinal tract, and skin by animals and humans. Inhalation exposure to HCN provides the most rapid route of entry (USEPA 1985).

B. Distribution

Once absorbed, cyanide is distributed rapidly by the blood throughout the body. Distribution patterns vary with the route of exposure (USEPA 1985).

[a] Based on USEPA (1985)

Reviews of Environmental Contamination and Toxicology, Vol. 107

Table 1. General information and properties of potassium cyanide

Property	Value
Chemical Abstracts Service (CAS) number	151-50-8
Chemical formula	KCN
Molecular weight	65.12
Physical state	Colorless liquid
Boiling point	—
Melting point	634.5°C
Density	1.553 (20°C)
Vapor pressure	—
Specific gravity	1.5
Water solubility	71.6 g/100 mL (20°C)
Log octanol/water partition coefficient	—
Odor threshold	—
Taste threshold	—
Air concentration conversion factor	2.707 mg/m^3 = 1 ppm

Source Towill et al. (1978), Weast (1980)

Table 2. General information and properties of hydrogen cyanide

Property	Value
Chemical Abstracts Service (CAS) number	74-90-8
Chemical formula	HCN
Molecular weight	27.03
Physical state	Colorless gas or liquid
Boiling point	25.70°C
Melting point	−13.24°C
Density	0.688 (20°C)
Vapor pressure	—
Specific gravity	0.7[a]
Water solubility	Miscible
Octanol/water partition coefficient	0.66
Odor threshold	1 mg/m^3[b]
Taste threshold	—
Air concentration conversion factor	0.90 mg/m^3 = 1 ppm

Source: Towill et al (1978), Weast (1980).

[a] NIOSH (1976) Criteria for a Recommended Standard Occupational exposure to hydrogen cyanide and cyanide salts (NaCN, KCN, and Ca[CN]$_2$) Washington, DC US Dept HEW

[b] Weiss (1986)

Table 3. General information and properties of sodium cyanide

Property	Value
Chemical Abstracts Service (CAS) number	143-33-9
Chemical formula	NaCN
Molecular weight	49.01
Physical state	Colorless solid
Boiling point	14.96°C
Melting point	563.7°C
Density	1.60
Vapor pressure	—
Specific gravity	1.6
Water solubility	48 g/100 mL (10°C)
Octanol/water partition coefficient	−0.44
Odor threshold	—
Taste threshold	—
Air concentration conversion factor	2.037 mg/m³ = 1 ppm

Dogs treated with KCN at single gavage doses equivalent to 20, 50, and 100 mg HCN (1 57, 4.42, or 8.40 mg HCN/kg) absorbed 72%, 24%, and 16 6% KCN, respectively (Gettler and Baine 1938). The dogs died within 155, 21, and 8 min, respectively, after dosing.

Source· Towill et al. (1978), Weast (1980).

High levels of cyanide were found in brains and livers of three human subjects who ingested fatal doses of cyanide (Gettler and Baine 1938).

In rabbits, intramuscular injection of HCN caused higher levels of cyanide in blood and tissues than did KCN administration (Ballantyne et al. 1972).

When radiolabeled KCN (5 mg/kg) was administered orally to rats over 24 hr, radioactivity in whole blood and plasma rapidly declined, while radioactivity in erythrocytes showed a small increase (Farooqui and Ahmed 1982). Most of the radioactivity in the erythrocytes (94.3%) was found in the hemolysate rather than the membranes. The heme fraction contained 70% of the radioactivity, while 14% to 25% and 5% to 10% was found in the globin and cell-membrane fractions, respectively.

Cyanide does not accumulate in blood and tissues following chronic exposure. Virtually no cyanide was found in the plasma or kidneys of rats treated with dietary concentrations of 100 and 300 ppm for 2 yr (Howard and Hanzal 1955). Low levels were found in erythrocytes (mean of 1.97 μg). Increased levels of thiocyanate were found in plasma (1,123 μg), erythrocytes (246 μg), liver (665 μg) and kidney (1,188 μg).

Yamamoto et al. (1982) found that rats treated orally (by gavage) with sodium cyanide (7 and 21 mg CN/kg) showed higher levels of cyanide in lungs and liver than in blood. After inhalation exposure to HCN at 356 and 1,180 ppm (392 and 1,298 mg/m^3), concentration in the lungs exceeded that in the blood.

C. Metabolism

Cyanide is detoxified by an intramitochondrial enzyme, rhodanese, which catalyzes the transfer of sulfur from a donor to cyanide to form the less toxic thiocyanate. Rhodanese is widely distributed throughout the body; high levels are found in the liver (USEPA 1985). Other minor detoxification pathways include spontaneous reaction with cystine to form 2-imino-4-thiozolidene carboxylic acid and with hydroxycobalamine to form cyanocobalamine, vitamin B_{12} (USEPA 1985).

D. Excretion

Cyanide is primary eliminated in the urine as thiocyanate, although some cyanide enters the metabolism of one-carbon compounds and is eliminated in expired air as CO_2. A small amount of HCN is eliminated in expired air (USEPA 1985).

Rats eliminated 80% of subcutaneously injected cyanide as thiocyanate in the urine, while 15% was eliminated as urinary 2-imino-thiozolidine carboxylic acid (Wood and Cooley 1956).

A man who had ingested 3 to 5 g KCN (at least 1.2 g HCN was present in the blood 3 hr later) eliminated a total of 237 mg thiocyanate in 72-hr urine (Liebowitz and Schwartz 1948).

III. Health Effects

The enzyme cytochrome oxidase enables cells to utilize oxygen. By inhibiting this enzyme, cyanide causes cellular anoxia (USEPA 1985).

A. Humans

Cases of acute exposure to cyanide by the oral route are commonly suicide attempts (NIOSH 1976). Signs of acute cyanide poisoning include rapid breathing, gasping, tremors, convulsions, and death (DiPalma 1971). Fatal oral doses of cyanide compounds range from 50 to 200 mg (0.7 to 2.9 mg CN^-/kg) (USEPA 1985). Within 20 min of ingestion of a fatal dose, clinical signs progress from hyperventilation, vomiting, unconsciousness, convulsions, rapid and irregular heart rate, gasping, vascular collapse, and cyanosis leading to death.

Although data regarding chronic oral exposure of humans to HCN, KCN, or NaCN are not available, there are a number of reports on the possible cyanide-related etiology of thyroid disorders and neuropathies in people living in certain tropical areas of Africa where the staple diet consists largely of cassava. Cassava contains a high level of the cyanogenic glycoside, linamarin, which releases cyanide on metabolism or acid hydrolysis in vivo (Makene and Wilson 1972; Osuntokun et al. 1969). The observed neuropathies are characterized by optic atrophy, nerve deafness, and spinal ataxia.

Case studies and epidemiological studies of case-hardeners, electroplaters, metal polishers, photographic material workers, and HCN fumigators have revealed effects typical of sublethal cyanide poisoning, including headache, dizziness, and thyroid enlargement (NIOSH 1976).

B. Animals

1. Short-Term Exposure. The acute oral LD_{50} for KCN was 10 mg/kg (4 mg CN^-/kg) in rats (Gaines 1969; Hayes 1967) and 8.5 mg KCN/kg (3.4 mg CN^-/kg) in mice (Sheehy and Way 1968). The LD_{50} of intraperitoneally administered NaCN for mice was 3.2 mg CN^-/kg (Kruszyna et al. 1982).

Mice administered 1 or 2 mg KCN/kg (0.4 or 0.8 mg CN^-/kg) intraperitoneally showed minimal or no effects, while 3 to 5 mg KCN/kg (1.2 to 2.0 mg CN^-/kg) resulted in signs of toxicity (convulsions, agitation) (Isom et al. 1982). A dose of 6 mg KCN/kg (2.4 mg CN^-/kg) resulted in 20% mortality.

Doses that are fatal to one species may be harmless to others. An oral dose of 3.8 mg KCN/kg (1.5 mg CN^-/kg) was fatal to a dog in 155 min (Gettler and Baine 1938), but a higher dose of 8 mg KCN/kg (3.2 mg CN^-/kg), equal to the LD_{50} in mice, had only minimal effects on guinea pigs (Basu 1983).

Rats tolerated higher doses of cyanide (80 mg CN^-/kg/d) when it was mixed in the diet (Kreutler et al. 1978) than when it was administered by gavage (4.0 mg CN^-/kg) (Ferguson 1962). Rats tolerated 25 daily doses of 10 mg KCN/kg (4 mg CN^-/kg) when the chemical was mixed in the diet; each of these doses was equal to the acute oral LD_{50} (Hayes 1967).

Rats tolerated higher oral doses of KCN (approximately 30 mg KCN/kg/d or 12 mg CN^-/kg/d for 21 d) when it was administered in drinking water (Palmer and Olson 1979) than when KCN was administered in a bolus (approximately 10 mg/kg KCN; 4.5 mg CN^-/kg) by gavage with water as the vehicle (Hayes 1967; Gaines 1969).

Rats receiving approximately 12 mg CN^-/kg/d for 21 d in drinking water had significantly increased liver weights compared with controls, while rats receiving approximately 8 mg CN^-/kg/d in the diet did not (Palmer and Olson 1979).

Beagle dogs consuming 3 mg CN^-/kg/d in the diet for 30 d showed no clinical signs of toxicity, effects on body weight, hematological changes, or histopathologic lesions (American Cyanamid Co. 1959).

2. Long-Term Exposure. As with short-term exposure, animals can better tolerate long-term exposure to higher doses of cyanide when it is administered in the diet or in drinking water (Hayes 1967; Palmer and Olson 1979) than as a bolus dose by gavage.

Pigs (sows) maintained on diets containing cyanide (30.3, 276.6, and 520.7 mg CN^-/kg diet) throughout gestation and lactation showed hyperplasia of kidney glomerular cells and accumulation of colloid and morphological changes in follicular cells of the thyroid (Tewe and Maner 1981a) (see Developmental Effects, below).

Weanling rats maintained on a diet containing 1,500 ppm KCN (approximately 30 mg CN^-/kg/d) for 11.5 mon had a significantly reduced body-weight gain, increased excretion of thiocyanate at 4 mon and at 11 mon, decreased plasma thyroxine levels, and decreased thyroxine secretion rates at 4 mon (Philbrick et al. 1979). The effects appeared to be greater in the animals on the vitamin B_{12}- and methionine-deficient diet. There were no definitive histopathologic lesions in the optic or central nervous system, thyroid, or sciatic tissues; however, vacuolization and myelin degeneration were observed in spinal cord sections.

Dogs receiving ≥ 0.27 mg CN^-/kg/d, administered in a capsule for 15 mon, had degenerative changes in ganglion cells of the central nervous system (Hertting et al. 1960). These effects may be due to the method of administration in capsule, similar to a bolus dose by gavage.

Rats maintained for 104 wk on diets that had been fumigated with HCN to give average dose levels of 76 mg/kg diet and 190 mg/kg diet (approximately 3.6 and 7.5 mg CN^-/kg bw/d for males and 4.6 and 10.8 mg CN^-/kg/d for females) showed no effects clinically or histologically (Howard and Hanzal 1955). The only effects noted were increased CN^- levels in red blood cells and increased thiocyanate levels in plasma, red blood cells, liver, and kidneys of animals from both treatment groups.

3. Reproductive Effects. No effects on reproductive performance were seen in pregnant rats fed 500 mg CN^-/kg diet (KCN) through gestation and lactation (Tewe and Maner 1981b).

4. Developmental Effects. Severe teratogenic effects were seen in hamsters administered cyanide by subcutaneously implanted osmotic minipumps that delivered cyanide at a rate of 3.3 to 3.4 mg CN^-/kg/hr (79.2 to 81.6 mg CN^-/kg/d) from day 6 through 9 of gestation (Doherty et al. 1982).

Piglets born to pigs maintained on diets containing cyanide (30.3, 276.6, and 520.7 mg CN^-/kg diet) throughout gestation and lactation showed reduced organ-to-body weight ratios of the thyroid, spleen, and heart in the high- and/or medium-dose groups relative to the low-dose group piglets (Tewe and Maner 1981a) (see Long-Term Exposure, above).

5. Mutagenicity. Potassium cyanide was not mutagenic in *Salmonella typhimurium* with or without metabolic activation (De Flora 1981). A study using HCN gas reported marginally mutagenic activity in *S. typhimurium* strain TA100 (Kushi et al. 1983). Addition of S-9 mix decreased the mutagenic activity.

Cyanide was not mutagenic in a modified rec assay in *Bacillus subtilis* (Karube et al. 1981).

6. Carcinogenicity. No information on the carcinogenicity of cyanides was located in the literature searched.

IV. Quantification of Toxicological Effects[a]

A. One-Day Health Advisory

The available data are insufficient to develop a One-Day HA for cyanide. It is recommended that the modified Drinking Water Equivalent Level (DWEL) calculated below of 0.22 mg/L (adjusted for a 10-kg child) be used as the One-Day HA.

B. Ten-Day Health Advisory

The modified DWEL of 0.22 mg/L (adjusted for a 10-kg child) is recommended to be used as the Ten-Day HA.

The study by Palmer and Olson (1979) was considered for use as the basis of the Ten-Day HA. This study reported a No-Observed-Adverse-Effect level (NOAEL) of 8 mg/kg/d. The NOAEL used in the modified DWEL is 10.8 mg/kg/d. Since the NOAELs from both studies are in general agreement, and greater confidence can be placed in the Howard and Hanzal (1955) study, the latter is selected to serve as the basis for the modified DWEL and Ten-Day HA.

C. Longer-Term Health Advisory

The available data are insufficient to develop Longer-Term HAs for cyanide. It is recommended that the DWEL of 0.77 mg/L used as the Longer-Term HA for a 70-kg adult and the modified DWEL of 0.22 mg/L (adjusted for a 10-kg child) be used as the Longer-Term HA for a 10-kg child.

D. Lifetime Health Advisory

The study of Howard and Hanzal (1955) has been selected to serve as the basis for the DWEL and Lifetime HA. Using the NOAEL from this study of 10.8 mg/kg/d, the DWEL and Lifetime HA are derived as follows:

[a] See Introduction for discussion of methodology.

Step 1: Determination of the Reference Dose (RfD)

$$\text{RfD} = \frac{(10.8\ \text{mg/kg/d})}{(100)(5)} = 0.022\ \text{mg/kg/d}^{a}$$

where 100 = uncertainty factor chosen in accordance with NAS/ODW guidelines for use with a NOAEL from an animal study; 5 = additional uncertainty factor selected to allow for possibly greater absorption of cyanide from water than from the diet.

Step 2: Determination of the Drinking Water Equivalent Level (DWEL)

$$\text{DWEL} = \frac{(0.022\ \text{mg/kg/d})(70\ \text{kg})}{(2\ \text{L/d})} = 0.77\ \text{mg/L}\ (770\ \mu\text{g/L})$$

Step 3: Determination of the Lifetime Health Advisory

The Lifetime Health Advisory is calculated as follows:

$$\text{Lifetime HA} = (0.77\ \text{mg/L})(20\%) = 0.154\ \text{mg/L}\ (154\ \mu\text{g/L})$$

E. Evaluation of Carcinogenic Potential

No information is available pertaining to the carcinogenicity of cyanides. The International Agency for Research on Cancer has not calculated the carcinogenic potential of cyanides. With the criteria described in USEPA's final guidelines for assessment of carcinogenic risk (USEPA 1986), cyanide is classified in Group D: not classified. This category is for agents with inadequate or no evidence of carcinogenicity in animals or humans.

V. Other Criteria, Guidance, and Standards

The proposed ambient water quality criterion is 3.77 mg/L, assuming that a 70-kg human consumes 2 L of water and 6.5 g of fish per day with a bioconcentration factor of 1.0 (USEPA 1982).

The U.S. Public Health Service (USPHS 1962) recommended that concentrations of cyanide in water supplies not exceed 0.2 mg/L. It also recommended that concentrations in drinking water be kept below 0.01 mg/L, since this level or lower can be achieved by proper treatment. The American Conference of Governmental Industrial Hygienists (ACGIH 1980) has recommended a Threshold Limit Value (TLV) for alkali cyanides in workroom air of 5 mg CN^-/m^3. National Institute for Occupational Safety and Health (NIOSH) has recommended a TLV of 5 mg/m^3 for CN^-, which was adopted by the Occupational Safety and Health Administration (OSHA 1981).

[a] NB: The RfD supports the observation (NIOSH 1976) that inhalation of 1 mg HCN/m^3 is without effect in humans.

VI. Analytical Methods

Cyanide levels are determined by volumetric titration or colorimetry (USEPA 1979). Cyanide is released from cyanide complexes as hydrocyanic acid (HCN) by means of a reflux-distillation operation and absorbed in a scrubber containing sodium hydroxide solution. The concentration of cyanide ion in the absorbing solution is then determined by volumetric titration or colorimetrically. The titration procedure uses a standard solution of silver nitrate and an indicator. Detection limit is 1 mg/L. In the colorimetric measurement, the cyanide is converted to cyanogen chloride, a reagent is added to form a colored complex, and the absorbance is measured. The detection limit is 20 μg/L.

VII. Treatment Technologies

Several treatment technologies for the removal of cyanide are available, although most reported treatments are wastewater applications. The treatment of high concentrations of cyanide and cyanide complexes in industrial waste streams and mine drainage runoff has been studied extensively, but only limited information is available on reductions of low cyanide levels in drinking water supplies.

The general treatment technologies that may be practical for reducing cyanide levels in drinking water include oxidation by chlorine or ozone, ion exchange, and reverse osmosis.

Oxidation by chlorine may be the cheapest and most practical method to remove cyanide from drinking water. In addition to cyanide removal, chlorine oxidation may also have secondary beneficial effects, such as disinfection of the water, oxidation of iron and manganese, and oxidation of taste- and odor-causing compounds. Practical experience in the wastewater industry and the laboratory indicates that chlorine oxidation removes 99% or more of the cyanide from the water (Gott 1978; Smith et al. 1980).

Oxidation by ozone may be used to destroy cyanide in water if the formation of trihalomethanes must be avoided. The use of ozone oxidation for cyanide is a relatively new technique. Laboratory and pilot studies by Cullivan (1978) indicate that complete destruction of cyanide in water requires an ozone-to-cyanide molar ratio of 1.3. These results were achieved with 5 to 15 mg/L of cyanide in the influent water. However, the results indicate that at cyanide concentrations of less than 5 mg/L, the rate of destruction is decreased.

Although reverse osmosis and ion exchange can reduce cyanide levels in the water, their application may not be economical if cyanide is the only contaminant to be removed. Experience by Moore (1976) and Trachtenberg and Murphy (1979) indicates that a well-designed ion exchange facility can remove over 99% of the cyanide present in the water. Pilot-plant studies performed by Rosehart (1973) in treating mine waters by reverse osmosis resulted in cyanide removal ranging from 28.7 to 81.6%.

References

ACGIH (1980) American Conference of Governmental Industrial Hygienists. Documentation of the threshold limit values for substances in workroom air, 4th Ed. Supplements through 1981. ACGIH, Cincinnati, OH, pp 109–110.

American Cyanamid Co (1959) Report on sodium cyanide: 30-day repeated feedings to dogs. Central Med Department. Report number 59-14. American Cyanamid, Wayne, NJ.

Ballantyne B, Bright J, Swanston DW, Williams P (1972) Toxicity and distribution of free cyanides given intramuscularly. Med Sci Law 12:209–219.

Basu TK (1983) High-dose ascorbic acid decreases detoxification of cyanide derived from amygdalin (laetrile): studies in guinea pigs. Can J Physiol Pharmacol 61(11):1426–1430.

Cullivan BM (1978) Industrial toxics oxidation: an ozone-chlorine comparison. Presented at the 33rd Purdue industrial waste conference. Purdue University, Lafayette, IN.

De Flora S (1981) Study of 106 organic and inorganic compounds in the *Salmonella*/ microsome test. Carcinogenesis 2(4):283–298.

DiPalma JR (ed) (1971) Noxious gases and vapors: I. Carbon monoxide, cyanides, methemoglobin and sulfhemoglobin. In: Drill's pharmacology in medicine. McGraw-Hill Book Co, New York, NY, pp 1189–1205.

Doherty PA, Ferm VH, Smith RP (1982) Congenital malformations induced by infusion of sodium cyanide in the golden hamster. Toxicol Appl Pharmacol 64:456–464.

Farooqui MYH, Ahmed AE (1982) Molecular interaction of acrylonitrile and potassium cyanide with rat blood. Chem Biol Interact 38:145–159.

Ferguson HC (1962) Dilution of dose and acute oral toxicity. Toxicol Appl Pharmacol 4:759–762.

Gaines TB (1969) Acute toxicity of pesticides. Toxicol Appl Pharmacol 14:515–534.

Gettler AO, Baine JO (1938) The toxicology of cyanide. Am J Med Sci 195:182–198.

Gott RD (1978) Development of waste water treatment at the Climax Mine. Mining Cong J 64(4):28–34.

Hayes WT (1967) The 90-dose LD_{50} and a chronicity factor as measurer of toxicity. Toxicol Appl Pharmacol 11:327–335.

Hertting G, Kraupp O, Schnetz E, Wieketich S (1960) Untersuchungen uber die Folgen einer chronischen Verabreichung akut toxischer Dosen von Natriumcyanid an Hunden. Acta Pharmacol Toxicol 17:27–43.

Howard JW, Hanzal RF (1955) Chronic toxicity for rats of food treated with hydrogen cyanide. J Agric Food Chem 3:325–329.

Isom GE, Burrows GE, Way JL (1982) Effect of oxygen on the antagonism of cyanide intoxication-cytochrome oxidase, *in vivo*. Toxicol Appl Pharmacol 65(2):250–256.

Karube I, Matsunaga T, Nakahara T, Suzuki S, Kata T (1981) Preliminary screening of mutagens with a microbial sensor. Anal Chem 53(7): 1024–1026.

Kreutler PA, Varbanov V, Goodman W, Olaya G, Stanbury JB (1978) Interactions of protein deficiency, cyanide and thiocyanate on thyroid function in neonatal and adult rats. Am J Clin Nutr 31: 282–289.

Kruszyna R, Kruszyna H, Smith RP (1982) Comparison of hydroxylamine, 4-dimethylaminophenol and nitrite protection against cyanide poisoning in mice. Arch Toxicol 49: 191–202.

Kushi A, Matsumoto T, Yoshida D (1983) Mutagen from the gaseous phase of protein pyrolyzate. Agric Biol Chem 47(9): 1979–1982.

Liebowitz D, Schwartz H (1948) Cyanide poisoning: report of a case with recovery. Am J Clin Pathol 18: 965–970.

Makene WJ, Wilson J (1972) Biochemical studies in Tanzanian patients with ataxic tropical neuropathy. J Neurol Neurosurg Psych 35: 31–33.

McCabe LJ, Symons JM, Lee RD, Robeck GG (1970) Survey of community water supply systems. J Am Waterworks Assoc 62: 670–687.

Moore FL (1976) An improved ion exchange resin method for removal and recovery of zinc cyanide and cyanide from electroplating wastes. J Environ Sci Hlth 7: 459–467.

NIOSH (1976) National Institute for Occupational Safety and Health. Criteria for a recommended standard...occupational exposure to hydrogen cyanide and cyanide salts [NaCN, KCN, and $Ca(CN)_2$]. NIOSH Publication no. 77-108. Department of Health, Education and Welfare (DHEW), U.S. Government Printing Office, Washington, DC.

OSHA (1981) Occupational Safety and Health Administration. General Industry OSHA Safety and Health Standards (29 CFR 1910). OSHA Publication no. 2206. U.S. Department of Labor, Washington, DC.

Osuntokun BO, Monekosso GL, Wilson J (1969) Relationship of a degenerative tropical neuropathy to diet, report of a field study. Brit Med J 1: 547–550.

Palmer IS, Olson OE (1979) Partial prevention by cyanide of selenium poisoning in rats. Biochem Biophys Res Commun 90(4): 1379–1386.

Philbrick DJ, Hopkins JB, Hill DC, Alexander JC, Thomson RG (1979) Effects of prolonged cyanide and thiocyanate feeding in rats. J Toxicol Environ Hlth 5: 579–592.

Rosehart RG (1973) Mine water purification by reverse osmosis. Can J Chem Eng 51(12): 788–789.

Sheehy M, Way JL (1968) Effect of oxygen on cyanide intoxication. III. Mithridate. J Pharmacol Exp Therap 161: 163–168.

Smith R, Siebert MS, Hattingh WHJ (1980) Removal of inorganic pollutants from waste water during reclamation for potable reuse. Water South Africa 6(2): 92–95.

Tewe OO, Maner JH (1981a) Performance and pathophysiological changes in pregnant pigs fed cassava diets containing different levels of cyanide. Res Vet Sci 30(2): 147–151.

Tewe OO, Maner JH (1981b) Long-term and carry-over effect of dietary inorganic cyanide (KCN) in the life cycle performance and metabolism of rats. Toxicol Appl Pharmacol 58(1): 1–7.

Towill LE, Arury JS, Whitfield BL, Lewis EB, Galyan EL, Hammone AS[a] (1978) Reviews of the environmental fate of pollutants: V. Cyanide. USEPA report no. EPA-600/1-78-027. NTIS[b] publication no. PB 289920. USEPA Health Effects Research Laboratory, Office of Research and Development (ORD), Cincinnati, OH.

Trachtenberg JJ, Murphy MA (1979) Removal of iron cyanide complexes from waste water utilizing an ion exchange process. Light Metals J 2:861–870.

USEPA[a] (1979) Method 335.2. Titrimetric; spectrophotometric. In: Methods for chemical analysis of water and wastes. Report no. EPA-600/4-79-020. March 1979. USEPA Environmental Monitoring and Support Laboratory, Cincinnati, OH.

USEPA[a] (1982) Ambient water quality criteria for cyanides, with errata for ambient water quality criteria documents dated June 9, 1981 (updated February 23, 1982). NTIS[b] publication no. PB 81-117483. USEPA Environmental Criteria and Assessment Office (ECAO), Cincinnati, OH.

USEPA[a] (1985) Health effects criteria document for cyanide. USEPA Office of Drinking Water, Washington, DC.

USEPA (1986) Final guidelines for carcinogen risk assessment. Fed Regis 51(185): 33992–34003. September 24, 1986.

USPHS (1962) U.S Public Health Service. Drinking water standards. Public Health Service (PHS) Publication no. 956. U.S. Government Printing Office, Washington, DC.

Weast RC (ed) (1980) CRC handbook of chemistry and physics. 61st Ed. CRC Press Inc, Boca Raton, FL. pp B-98, B-133, B-147.

Weiss G (ed) (1986) Hazardous chemical's Data Book, 2nd Ed. Noyes Data Corporation, Park Ridge, NJ. p 572.

Wood JL, Cooley SL (1956) Detoxication of cyanide by cystine. J Biol Chem 218:449–457.

Yamamoto K, Yamamoto Y, Hattori H, Samori T (1982) Effects of routes of administration on the cyanide concentration distribution in the various organs of cyanide-intoxicated rats. Tohoku J Exp Med 131:73–78.

---

[a] USEPA documents can be obtained by writing to the appropriate office within the U.S. Environmental Protection Agency, 401 M Street, S W., Washington, DC 20460. They may also be available by contacting the National Technical Information Service, 5285 Port Royal Road, Springfield, VA 22161 (Telephone· 1-800-336-4700).

[b] Documents with an NTIS publication number are available from the National Technical Information Service, 5285 Port Royal Road, Springfield, VA 22161 (Telephone 1-800-336-4700).

Epichlorohydrin[a]

I. General Information

Epichlorohydrin is also known as 1-chloro-2,3-epoxypropane, 3-chlorol-1,2-epoxypropane, chloromethyl oxirane, 2-(chloromethyl)oxirane, and chloropropylene oxide. It is used in the manufacture of epoxide resins, surface active agents, pharmaceuticals, and agricultural chemicals (Verschueren 1983). Its structure and properties are shown in Table 1.

Total U.S. production of epichlorohydrin in 1982 was approximately 350 million lb (158 million kg). Although epichlorohydrin reportedly hydrolyzes readily in aqueous solution (hydrolysis half-life of 8.2 d at 20°C and pH 7) to water-soluble alcohols, its use in water-treatment resins and coatings makes exposure possible (Mabey and Mill 1978).

No information has been located in either state or federal surveys to confirm the presence or absence of epichlorohydrin in drinking water.

II. Pharmacokinetics

A. Absorption

Epichlorohydrin is readily absorbed after oral, inhalation, or dermal exposures (USEPA 1985a). Gingell et al. (1985) assessed its pharmacokinetics and metabolism. Male Fischer-344 rats were treated once (6 mg/kg by gavage) with [2-^{14}C]epichlorohydrin (98% pure) in water and sacrificed after 3 d. Ready absorption was shown by an initial elimination half-life of 2 hr and total excreta recovery of 91.61% of the radiolabelled chemical.

Smith et al. (1979) reported extensive absorption of epichlorohydrin in water by male Fischer-344 rats (weighing 190 to 220 g) after a single gavage exposure. Based on excretion data, the extent of absorption, approximately 100% within 72 hr after administration, appeared to be similar after doses of either 1 or 100 mg/kg.

Smith et al. (1979) indicated epichlorohydrin was absorbed readily by male Fischer-344 rats after 6-hr exposure to atmospheres containing 1 or 100 ppm epichlorohydrin (approximately 3.78 or 378 mg/m^3). Uptake rates of 15.48

[a] Based on USEPA (1985a)

Table 1. General information and properties of epichlorohydrin

Property	Value
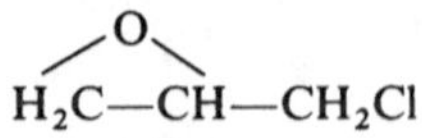	
Chemical Abstracts Service (CAS) number	106-89-8
Chemical formula	C_3H_5ClO
Molecular weight	92.53
Physical state	Colorless liquid
Boiling point	116.1°C
Melting point	−57.2°C
Density	1.18 (20°C)
Vapor pressure	12 mm Hg (20°C)
Specific gravity	1.18 (20°C)
Water solubility	66 g/L (20°C)
Log octanol/water partition coefficient	0.26
Odor threshold	0.5–1.0 mg/L, 3 mg/L[a]
Taste threshold	—
Air concentration conversion factor	3.78 mg/m^3 = 1 ppm

Source USEPA (1985a)
[a] Amoore and Hautala (183)

and 1,394 μg/hr were calculated for exposures to 1 and 100 ppm in air, respectively. The investigators stated these exposures corresponded to doses of 0.37 and 33 mg/kg.

The toxicity study of Kremneva and Tolgskaya (1961) indicated epichlorohydrin is absorbed after dermal exposure. When the tails of mice were immersed in epichlorohydrin for a single exposure of 1 hr or for repeated exposures of 20 to 30 min/d for 2 to 3 d consecutively, toxic signs and death were observed within 3 d.

B. Distribution

Gingell et al. (1985) reported that 8 to 9% of ^{14}C-labelled epichlorohydrin was distributed to tissues, with the highest levels (specific activity, disintegrations/min × 10^{-3}/g tissue; wet weight) in liver (177.5; 2.82% of dose), kidney (127.1; 0.41% of dose), and forestomach (81.6; 0.03% of dose).

Smith et al. (1979) compared the distribution of [1,3-^{14}C]epichlorohydrin in male Fischer-344 rats following oral (100 mg/kg) or inhalation (100 ppm for 6 hr) exposure. At 3 hr after exposure in the oral study and at the termination of inhalation exposure, the plasma levels of radioactivity were 36.1 and 18.3 mg/g, respectively. Concentrations in tissues were expressed as μg equivalents of epichlorohydrin/g tissue. After oral treatment, the greatest concentrations were in the stomach, followed by the intestine, kidney, liver, pancreas, and lung. After inhalation exposure, the highest levels were in the nasal turbinates, followed by the intestine, liver, and kidneys.

C. Metabolism

Gingell et al. (1985) concluded that the initial elimination half-life of 2 hr measured in their study indicated rapid metabolism. Main urinary metabolites were *N*-acetyl-*S*-(3-chloro-2-hydroxypropyl)-*L*-cysteine and α-chlorohydrin, which represented 36% and 4% respectively of the delivered doses. One major metabolite and four minor metabolites were identified in urine. These investigators stated that the presence of the two dominant urinary metabolites is consistent with initial metabolic reactions consisting of conjugation of the epoxide with glutathione and hydration of the epoxide.

Smith et al. (1979) administered [1,3-^{14}C]epichlorohydrin to male Fischer 344 rats as single oral doses of 1 or 100 mg/kg or as 6-hr inhalation exposures to 1 or 100 ppm (approximately 3.78 or 378 mg/m^3). Urinary metabolites were separated by ion-exclusion chromatography. Seven radioactive peaks were found in urine after oral dosing and six radioactive peaks after inhalation exposure, but none corresponded to epichlorohydrin. The authors noted the patterns of urinary metabolite excretion were similar after oral or inhalational dosing; metabolites were not identified.

Epichlorohydrin has two electrophilic centers and may bind to cellular nucleophiles. It is also a substrate for epoxide hydratase resulting in the formation of α-chlorohydrin, which may be oxidized to oxalic acid, converted to glycidol, or phosphorylated to 3-chloroglycerophosphate (USEPA 1985a). Gingell et al. (1985) did not find oxalic acid as a metabolite, however.

Rossi et al. (1983) found epichlorohydrin rapidly disappeared from the blood of CD-1 mice. The half-life was approximately 5 min, with α-chlorohydrin appearing as epichlorohydrin levels decreased. α-Chlorohydrin, however, had a much longer elimination half-life (50 to 60 min).

D. Excretion

In the study by Gingell et al. (1985), the half-life for initial elimination of ^{14}C in both urine and exhaled air was about 2 hr. Approximately 38% of the radioactive dose was exhaled as carbon dioxide (CO_2), 50% was excreted as urinary metabolites, and 39% was eliminated in feces.

Smith et al. (1979) administered [1,3-^{14}C]epichlorohydrin by single gavage doses of 1 or 100 mg/kg to groups of four male Fischer 344 rats. In parallel experiments, three rats each were exposed (head only) to atmospheres containing 1 and 100 ppm epichlorohydrin for 6 hr. The rates and routes of excretion were essentially unaffected by either the dose or route. Urine, the major route of excretion, accounted for 46% to 54% of the dose. An additional 25% to 42% was recovered as $^{14}CO_2$ in expired air. Only 3% to 6% of the dose was recovered in feces. Excretion was biphasic, with an initial rapid phase that dominated the first 24 hr after exposure and a slower second phase that was dominant after 24 hr. The calculated half-lives for elimination from the plasma were 1 to 2 hr and 26 to 27 hr for the fast and slow phases, respectively.

III. Health Effects

A. Humans

In humans, acute effects have been reported after both dermal and inhalation exposures. Dermal exposure produced predominantly local irritation effects, but inhalation produced significant systemic effects, including hepatic and renal toxicity. In a case report of a worker exposed to epichlorohydrin vapor, systemic effects were evident for at least 2 yr after exposure (USEPA 1985a). Chronic exposure has been associated with chromosomal and chromatid breaks, decreased hemoglobin concentration, decreased erythrocyte counts, and decreased leukocyte counts. Increased (not statistically significant) deaths from lung cancer have been reported in workers sequentially exposed to isopropyl alcohol and epichlorohydrin (USEPA, 1985a). No effects on reproductive function have been detected.

B. Animals

1. Short-Term Exposure. Epichlorohydrin is acutely toxic after oral, percutaneous, subcutaneous, or respiratory exposure and produces similar symptoms in all types of exposures. At the site of application, it is a strong irritant. The major acute systemic effects occur in the central nervous system (CNS); deaths are caused by depression of the respiratory center. The major internal organs affected are the lungs, liver, and kidneys (USEPA 1985a).

Oral doses as low as 10 mg/kg for 5 d/wk for 2 wk resulted in decreased ($p < 0.05$) erythrocyte counts in male rats and decreased ($p < 0.01$) kidney-to-body weight ratios in females (Van Esch 1981). Similar exposures to 40 or 80 mg/kg resulted in degenerative changes in the kidneys of male and female rats.

2. Long-Term Exposure. Epichlorohydrin given in drinking water at 375, 750, and 1,500 ppm (18, 39, and 89 mg/kg/d) to male Wistar rats for 81 wk induced forestomach hyperplasia and decreased body weight at all doses (Konishi et al. 1980).

With gavage administration of epichlorohydrin in water at doses of 2 and 10 mg/kg, 5 d/wk, for 104 wk, stomach hyperplasia and a dose-related decrease in white blood cells were observed in male and female Wistar rats (Van Esch 1982).

Inhalation exposure of Fischer 344 rats, Sprague–Dawley rats, B6C3F1 mice and New Zealand rabbits to epichlorohydrin at 19 mg/m^3 for 90 d was without observable effect. Higher exposure levels induced nasal irritation, eye irritation, kidney lesions, and respiratory tract lesions (John et al. 1983; Quast et al. 1979).

Lifetime inhalation exposure of male Sprague–Dawley rats to 38 and 114 mg/m^3 produced kidney lesions (Laskin et al. 1980).

3. Reproductive Effects. Male and female Wistar rats were given epichlorohydrin in water starting 10 d before mating and continuing for 3 mon (Van Esch 1981). A dose of 2 mg/kg was ineffective. A 10-mg/kg dose reduced fertility, and cross-mating with untreated rats attributed the antifertility effect to males. Sterility of male rats given epichlorohydrin orally also was observed by Hahn (1970) and Cooper et al. (1974), with gavage doses of 15 mg/kg and higher for 15 or 5 d; however, these investigators showed the effect to be reversible.

Exposure of male rats to epichlorohydrin by inhalation at levels greater than 19 mg/m^3 for 10 wk resulted in reversible sterility, and fertility of male rabbits was unaffected by inhalation exposure levels as high as 189 mg/m^3 (John et al. 1983).

4. Developmental Effects. Epichlorohydrin was not teratogenic when given by gavage in cottonseed oil to pregnant CD rats and CD-1 mice on d 6 through 15 of gestation (Marks et al. 1982). Doses above 40 mg/kg were maternally toxic (reduced body weight, increased liver weight, death) in rats. Doses above 80 mg/kg were maternally toxic (increased liver weight and death) and fetotoxic (reduced body weight) in mice.

Inhalation exposures of pregnant Sprague–Dawley rats and New Zealand rabbits to 9.5 and 95 mg/m^3 during gestational d 6 through 15 (rats) and 6 through 18 (rabbits) were neither teratogenic nor fetotoxic. Pregnant rats exposed to 95 mg/m^3 weighed less than controls (Pilny et al. 1979).

5. Mutagenicity. Epichlorohydrin is a mutagen in several systems (USEPA 1985a). It is a potent inducer of base-pair, substitution-type mutations in prokaryotic systems. Incubation with mammalian liver homogenates results in a marked reduction in mutation frequency. It induced gene mutations and probably chromosomal aberrations in mouse lymphoma cell cultures (Moore-Brown and Clive 1979) and clastogenesis in human lymphocytes in vitro (Norppa et al. 1981) but not in rat liver cell cultures (Dean and Hodson-Walter 1979). Epichlorohydrin induced sister-chromatid exchange in cultured human lymphocytes (Carbone et al. 1981; Norppa et al. 1981;

White 1980). Examination of occupationally exposed workers indicated that chromosomal aberrations also occurred in vivo (Kucerova et al. 1977; Picciano 1979a,b; Sram et al. 1976).

In in vivo studies, epichlorohydrin treatment resulted in an increased incidence of sex-linked recessive lethals in *Drosophila melanogaster* when the chemical was administered by injection but not when it was incorporated in food (Knapp et al. 1982; Wurgler and Graf 1981). In other in vivo studies, it produced negative results in the mouse dominant lethal assay (Epstein et al. 1972; Sram et al. 1976) and the mouse micronucleus assay (Kirkhart 1981; Tsuchimoto and Matter 1981). Clastogenic effects in bone marrow cells in vivo were found in mice (Sram et al. 1976) but not in rats (Dabney et al. 1979).

6. Carcinogenicity. Epichlorohydrin is carcinogenic at the site of administration. Administration of 375, 750, and 1,500 ppm epichlorohydrin in drinking water (equivalent to 18, 39, and 89 mg/kg/d, based on data provided by the authors (i.e., total doses of 5.0, 8.9, and 15.1 g/rat during 81 wk of treatment divided by body weight) to male Wistar rats for 81 wk resulted in forestomach hyperplasia at all doses and papillomas and carcinomas of the forestomach at the two highest doses (Konishi et al. 1981).

Lifetime gavage treatment of male and female Wistar rats with aqueous epichlorohydrin solution at doses of 2 and 10 mg/kg induced papillomas and carcinomas of the forestomach (Wester et al. 1985; Van Esch 1982).

Laskin et al. (1980) found nasal carcinomas in male Sprague-Dawley rats exposed to 378 mg/m^3 by inhalation 6 hr/d, 5 d/wk for 6 wk followed by lifetime observation.

Subcutaneous injection in ICR/Ha Swiss mice induced local sarcomas; epichlorohydrin was effective as an initiator but not as a complete carcinogen on skin of ICR/Ha Swiss mice (Van Duuren et al. 1972, 1974).

IV. Quantification of Toxicological Effects[a]

A. One-Day Health Advisory

Because appropriate data for calculation of a One-Day HA are not available, the Ten-Day HA (0.14 mg/L) is recommended for use as the One-Day HA.

B. Ten-Day Health Advisory

The reproductive toxicity study by Van Esch (1981) can be used to derive the Ten-Day HA. In this study, male and female rats were given epichlorohydrin by gavage 5 d/wk at doses of 0, 2, or 10 mg/kg. Exposure was started 10 d before mating and continued until the F_{1b} generation was produced. The

[a] See Introduction for discussion of methodology

fertility index at the first mating was reduced in the high-dose group but not in the low-dose group. The study of Hahn (1970), which reported infertility in male rats exposed by gavage at 15 mg/kg/d for 12 d, supports an assumption that at least a portion of the reduced fertility index observed by Van Esch (1981) was the result of infertility in the males associated with the 10-d exposure before mating. In this study, 2 mg/kg was a No-Observed-Adverse-Effect Level (NOAEL) for reproductive effects and is appropriate for use in deriving the Ten-Day HA.

Using the NOAEL of 2 mg/kg/d, the Ten-Day HA for a 10-kg child is calculated as follows:

$$\text{Ten-Day HA} = \frac{(2\ \text{mg/kg/d})(10\ \text{kg})(5)}{(100)(1\ \text{L/d})(7)} = 0.14\ \text{mg/L}\ (140\ \mu\text{g/L})$$

where 5/7 = conversion of dose to represent continuous exposure (7 d/wk).

Although the antifertility effect in male rats in the Van Esch (1981) study pertains to men as a specific sensitive subpopulation for this effect, this study is preferred for calculating a Ten-Day HA for the general population because of its design (oral short-term exposure) and its demonstration of no-effect and effect levels. In addition, the 2 mg/kg NOAEL in the Van Esch (1981) study appeared consistent with the dose responses in the overall Van Esch (1981) work in which both systemic and reproductive effects were found with 10-d oral exposures to 10 mg/kg.

C. Longer-Term Health Advisory

Data are insufficient for calculating a Longer-Term HA. The Drinking Water Equivalent Level (DWEL) of 0.07 mg/L is recommended as a conservative estimate of the Longer-Term HA.

D. Lifetime Health Advisory

The reviewed studies in which the effects of long-term exposure to epichlorohydrin were investigated are Laskin et al. 1980; Konishi et al. 1980, also reported by Kawabata 1981; and Wester et al. 1985, also reported by Van Esch 1982. The Laskin et al. (1980) study was selected as the most appropriate from which to derive the DWEL. Forestomach hyperplasia in all three treatment groups and papillomas and carcinomas of the forestomach in the two highest dose groups were found in the study by Konishi et al. (1980). Because the hyperplasia could be considered a pre-neoplastic effect and the progression of forestomach lesions beyond the 81-wk duration of this study is uncertain, it would be questionable to use this effect in the low-dose group (18 mg/kg/d) for calculating a DWEL for drinking-water exposure. Dose–response data for toxicity and carcinogenicity in the Konishi et al. (1980) drinking water study are preferred over the bolus gavage dosing study by Wester et al. (1985).

Use of the estimated 2.16 mg/kg/d dose in the Laskin et al. (1980) study was concluded to be consistent with the dose–response indicated by the Konishi et al. (1980) study. The Lowest-Observed-Adverse-Effect Level (LOAEL) is based on renal damage at 2.16 mg/kg/d as estimated from data in the Laskin et al. (1980) study. This LOAEL was used to derive a DWEL. In addition, carcinogenic effects were not apparent at the LOAEL in the Laskin et al. (1980) study. Using this LOAEL, the DWEL is derived as follows:

Step 1: Conversion of Inhalation Exposure to Oral Exposure

Applying the 38 mg/m^3 inhalation LOAEL in the Laskin et al. (1980) study and the assumptions in USEPA (1985a) for converting inhalation exposure to oral exposure for the rat, the estimated oral dose would be:

$$\text{Total Absorbed Dose} = \frac{(38\ \text{mg/m}^3)(0.0093\ \text{m}^3/\text{hr})(6\ \text{hr/d})(5)(0.5)}{(0.35\ \text{kg})(7)}$$

$$= 2.16\ \text{mg/kg/d}$$

where: 38 mg/m^3 = LOAEL; 0.0093 m^3 = amount of air breathed by a rat/hr; 6 hr/d = a 6-hr exposure each d; 5/7 = factor to convert from a 5 d/wk to 7 d/wk exposure; 0.5 = assumed inhalation absorption factor; 0.35 kg = assumed weight of a rat.

Step 2: Determination of the Reference Dose (RfD)

$$\text{RfD} = \frac{(2.16\ \text{mg/kg/d})}{(1{,}000)} = 0.002\ \text{mg/kg/d}\ (2\ \mu\text{g/kg/d})$$

Step 3: Determination of the Lifetime Health Advisory

$$\text{DWEL} = \frac{(0.002\ \text{mg/kg/d})(70\ \text{kg})}{(2\ \text{L/d})} = 0.07\ \text{mg/L}\ (70\ \mu\text{g/L})$$

A Lifetime Health Advisory is not recommended because epichlorohydrin is classified in Group B2: probable human carcinogen (see Evaluation of Carcinogenic Potential below). For comparison purposes, the cancer risk associated with the DWEL was determined.

The estimated excess cancer risk associated with lifetime exposure to drinking water containing epichlorohydrin at 70 μg/L is approximately 2×10^{-5}. This estimate represents the upper 95% confidence limit from extrapolations prepared by the USEPA's Carcinogen Assessment Group (CAG) using the linearized, multistage model. The actual risk is unlikely to exceed this value.

E. Organoleptic Properties

A reported threshold for odor perception of epichlorohydrin is 0.5 to 1.0 mg/L, and 0.1 mg/L was cited by the National Academy of Sciences (NAS

1980) as the threshold for its irritant action. Amoore and Hautala (1983) reported an odor threshold of 3 mg/L.

F. Evaluation of Carcinogenic Potential

Applying the criteria described in USEPA guidelines for assessment of carcinogenic risk (USEPA 1986), epichlorohydrin may be classified in Group B2: probable human carcinogen. This group is for agents for which there is inadequate evidence of carcinogenicity from human studies and sufficient evidence from animal studies.

Konishi et al. (1980) provided appropriate data for a quantitative risk assessment based on the relevant route of exposure and the observed dose-response pattern. Using the calculated $q_1{}^*$ of 9.9×10^{-3} $(mg/kg/d)^{-1}$, the 95% upper-limit lifetime dose associated with a 10^{-5} risk level is 70.7 μg/d. Assuming an average water consumption of 2 L/d, this risk level corresponds to a drinking water concentration of 35.4 μg/L. Corresponding drinking water levels for risks of 10^{-6} and 10^{-4} are 3.54 and 354 μg/L, respectively.

Maximum likelihood estimates (MLE) as well as 95% upper limits (UL) of cancer risks by the multistage model have been calculated (USEPA 1984). At 10 μg/L, cancer risk estimates are 1.4×10^{-17} (MLE) and 2.8×10^{-6} (UL); at 100 μg/L, cancer risk estimates are 2.6×10^{-14} (MLE) and 2.8×10^{-5} (UL).

The USEPA CAG estimated cancer risks with other models besides the multistage (USEPA 1984). As an example, 10 μg/L lifetime exposure was associated with additional risks (95% upper confidence limit) of 2.8×10^{-5} by the multistage, 3.4×10^{-5} by the one-hit, 0 by the Weibull, and 0 by the log-probit model. In the interest of consistency of approach and to provide an upper bound on potential cancer risk, the USEPA recommends use of the linearized multistage approach.

V. Other Criteria, Guidance, and Standards

The NAS (1980) established Suggested-No-Adverse-Response Levels (SNARLs) for 1-d or 7-d exposures to epichlorohydrin at 0.84 and 0.53 mg/L, respectively. NAS has not calculated an Acceptable Daily Intake or a cancer risk.

The American Conference of Governmental Industrial Hygienists recommended a Threshold Limit Value of 2 ppm (10 mg/m^3) (ACGIH 1982). Current Occupational Safety and Health Administration (OSHA) standards allow a Time-Weighted Average occupational exposure of 19 mg/m^3 (OSHA 1985). However, OSHA is currently considering lowering this value to 0.5 ppm (2 mg/m^3) with a ceiling value of 15 ppm (60 mg/m^3) for 15 min. Occupational standards in other countries range from 0.26 ppm in the USSR and Czechoslovakia to 3.6 ppm in the Federal Republic of Germany (Sram et al. 1980).

Epichlorohydrin has not been regulated under the Safe Drinking Water Act; however, discharge of >1,000 lb (454 kg) into navigable waters is prohibited under the Clean Water Act (USEPA 1985e). It is also classified as a "hazardous waste" by the USEPA, and quantities exceeding 100 kg must be disposed of in a special landfill (USEPA 1985f,g).

The proposed Recommended Maximum Contamination Level (RMCL) by the USEPA Office of Drinking Water is zero (USEPA 1985b).

VI. Analytical Methods

No standardized method exists for the determination of epichlorohydrin in drinking water samples. However, it may be determined by a purge-and-trap gas chromatographic (GC) and mass spectrometric (MS) procedure used for the determination of volatile organic compounds in water (USEPA 1985c). This method calls for the bubbling of an inert gas through the sample and trapping epichlorohydrin on an adsorbent. The absorbent is heated to drive off epichlorohydrin onto a GC column. The GC is temperature programmed to separate the method analytes, which are then detected by MS.

VII. Treatment Technologies

No data are available on the removal of epichlorohydrin from potable water by any treatment technique (ESE 1984; USEPA 1985d).

The amenability of epichlorohydrin to removal by conventional treatment or by adsorption is not known. The Henry's Law constant was estimated to be 2.44×10^{-5} atm $\times$ m^3/mole (ESE 1984). This value suggests that aeration is unlikely to be a successful removal technique. It also has been concluded that epichlorohydrin would not be removed from water by ozone oxidation (USEPA 1985d).

References

ACGIH (1982) American Conference of Governmental Industrial Hygienists TLVs—threshold limit values for chemical substances and physical agents in the workroom environment. ACGIH, Cincinnati, OH.

Amoore JE, Hautala E (1983) Odor as an aid to chemical safety: odor threshold compounds with threshold unit values and volatilities for 214 industrial chemicals in air and water dilution. J Appl Toxicol 3:272–290.

Carbone P, Barbata G, Margiotta G, Tomasino A, Granata G (1981) Low epichlorohydrin concentrations induce sister chromatid exchanges in human lymphocytes *in vitro*. Caryologia 34(3):261–266.

Cooper ER, Jones AR, Jackson H (1974) Effects of alpha-chlorohydrin and related compounds on the reproductive organs and fertility of the male rat. J Reprod Fert 39(2):379–386.

Dabney BJ, Johnston RV, Quast JF, Park CN (1979) Epichlorohydrin—subchronic studies. III. Cytogenetic evaluation of bone marrow cells from rats exposed by inhalation to epichlorohydrin for four weeks. ICPEMC (International Commission for Protection Against Environmental Mutagens and Carcinogens) Document no. 128, 15 pp.

Dean BJ, Hodson-Walker G (1979) An in vitro chromosome assay using cultured rat-liver cells. Mutat Res 64:329–337.

ESE (1984) Environmental Science and Engineering. Review of treatability data for removal of twenty-five synthetic organic chemicals from drinking water. USEPA Office of Drinking Water, Washington, DC.

Epstein SS, Arnold E, Andrea J, Bass W, Bishop Y (1972) Detection of chemical mutagens by the dominant lethal assay in the mouse. Toxicol Appl Pharmacol 23:288–325.

Gingell R, Mitschke HR, Dzidic I, Beatty PW, Sarvin VL, Page AC (1985) Disposition and metabolism of [2-^{14}C]epichlorohydrin after oral administration to rats. Drug Metab Dispos 13:333–341.

Hahn JD (1970) Post-testicular antifertility effects of epichlorohydrin and 2,3-epoxypropanol. Nature (London) 226:87.

IARC (1982) International Agency for Research on Cancer. IARC monographs on the evaluation of the carcinogenic risk of chemicals to humans. Chemical industrial processes and industries associated with cancer in humans. Supplement 4, pp 122–123. IARC, Lyon, France.

John JA, Quast JF, Murray FJ, Calhoun LS, Staples RE (1983) Inhalation toxicity of epichlorohydrin: effects on fertility in rats and rabbits. Toxicol Appl Pharmacol 68.415–423.

Kawabata A (1981) Studies on the carcinogenic activity of epichlorohydrin by oral administration in male Wistar rats. J Nara Med Assoc 32:270–280.

Kirkhart B (1981) Micronucleus test on 21 compounds. In: deSerres FJ and Ashby J (eds) Evaluation of short-term tests for carcinogens. Elsevier/North Holland Biomedical Press, Amsterdam, pp 698–704.

Knapp AGAC, Voodg CE, Kramers PGN (1982) Comparison of the mutagenic potency of 2-chloroethanol, 2-bromoethanol, 1,2-epoxybutane epichlorohydrin and glycidaldehyde in *Klebsiella pneumoniae*, *Drosophila melanogaster* and L5178Y mouse lymphoma cells. Mutat Res 101:199–208.

Konishi T, Kawabata A, Denda A, Ikeda T, Katada H, Maruyama H, Higashiguchi R (1980) Forestomach tumors induced by orally administered epichlorohydrin in male Wistar rats. Gann 71:922–923.

Kremneva SN, Tolgskaya MS (1961) Toxicology of epichlorohydrin. Toksikol Nov Prom Khim Veschestv 2:28–41.

Kucerova M, Zhurkov VS, Polwkova Z, Ivanova JE (1977) Mutagenic effect of epichlorohydrin. II. Analysis of chromosomal aberrations in lymphocytes of persons occupationally exposed to epichlorohydrin. Mutat Res 48:355–360.

Laskin S, Sellakumar AR, Kuschner M, Nelson N, LaMendola S, Ruch GM, Katz GV, Dulak NC, Albert RE (1980) Inhalation carcinogenicity of epichlorohydrin in non-inbred Sprague–Dawley rats. J Natl Cancer Inst 65(4):751–758.

Mabey W, Mill T (1978) Critical review of hydrolysis of organic compounds in water under environmental conditions. J Phys Chem Ref Data 7:385–415.

Marks TA, Gerling FS, Staples RE (1982) Teratogenic evaluation of epichlorohydrin in the mouse and rat and glycidol in the mouse. J Toxicol Environ Hlth 9:87–96.

Moore-Brown MM and Clive D (1979) The L5178Y/TK$^{+/-}$ mutagen assay system: *in situ* results. Banbury Report 2:71–88.

NAS (1980) National Academy of Sciences. Drinking water and health. Vol 3. National Academy Press, Washington, DC. pp 111–124.

Norppa H, Hemminki K, Sorsa M, Vainio H (1981) Effect of monosubstituted epoxides on chromosome aberrations and SCE in cultured human lymphocytes. Mutat Res 91:243–250.

OSHA (1985) Occupational Safety and Health Administration. Code of Federal Regulations. Title 29—Labor. Part 1910—Occupational safety and health standards. Subpart 2—Toxic and hazardous substances. Section 1910.1000—Air contaminants. U.S. Government Printing Office, Washington, DC, p. 656. July 1, 1985.

Picciano D (1979a) Cytogenic investigation of occupational exposure to epichlorohydrin. Mutat Res 66:169–173.

Picciano D (1979b) Faulty experimental design and underutilization of cytogenetic data. Benzene and epichlorohydrin. Ann NY Acad Sci 329:321–327.

Pilny MK, Lederer TS, Murray JS, et al. (1979) Epichlorohydrin subchronic studies. IV. The effects of maternally inhaled epichlorohydrin on rat and rabbit embryonal and fetal development. Unpublished report. Toxicol Res Lab Health Environ Sci, Dow Chemical USA, Midland, MI.

Quast JF, Henck JW, Pastma BJ, Scheutz DJ, McKenna MJ (1979) Epichlorohydrin subchronic studies. I. A 90-day inhalation study in laboratory rodents (Fischer-344 rats, Sprague-Dawley rats, and B6C3F$_1$ mice). Dow Chemical USA, Midland, MI, 166 pp.

Rossi AM, Migliore D, Lascialfari D, Sbrana I, Loprieno N (1983) Genotoxicity, metabolism, and blood kinetics of epichlorohydrin in mice. Mutat Res 118:213–226.

Smith FA, Langvardt PW, Young JD (1979) Pharmacokinetics of epichlorohydrin (EPI) administered to rats by gavage or inhalation. Dow Chemical USA, Toxicol Res Lab, Midland, MI, 51 pp.

Sram RJ, Cerna M, Kucerova M (1976) The genetic risk of epichlorohydrin as related to the occupational exposure. Biol Zbl 95:451–462.

Sram RJ, Zudova Z, Kuleshov NP (1980) Cytogenic analysis of peripheral lymphocytes in workers occupationally exposed to epichlorohydrin. Mutat Res 70:115–120.

Tsuchimoto T, Matter BE (1981) Activity of coded compounds in the micronucleus test. In: Seeres FJ and Ashby J (eds) Evaluation of short-term tests for carcinogens. Elsevier North-Holland Biomedical Press, Amsterdam, pp 705–711.

USEPA[a] (1984) Health assessment document for epichlorohydrin. Final report. Report no. EPA-600/8-83-032F. USEPA Office of Health and Environmental Assessment, Washington, DC.

[a] USEPA documents can be obtained by writing to the appropriate office within the U.S. Environmental Protection Agency, 401 M Street, S W., Washington, DC 20460. They may also be available by contacting the National Technical Information Service, 5285 Port Royal Road, Springfield, VA 22161 (Telephone. 1-800-336-4700)

USEPA[a] (1985a) Drinking water criteria document for epichlorohydrin. Final Draft. ECAD-CFN-413. USEPA Office of Drinking Water, Washington, DC.

USEPA (1985b) National primary drinking water regulations; Synthetic organic chemicals, inorganic chemicals and microorganisms; Proposed rule. Fed Regis 50(219):4693–4702, November 13.

USEPA[a] (1985c) Method 524.1. Volatile organic compounds in water by purge and trap gas chromatography/mass spectrometry. USEPA Environmental Monitoring and Support Laboratory (EMSL), Cincinnati, OH.

USEPA[a] (1985d) Technologies and costs for removal of organic chemicals from potable water supplies (Draft). USEPA Science and Technology Branch, Criteria and Standards Division, Office of Drinking Water, Washington, DC.

USEPA[a] (1985e) Designation of hazardous substance. Code of Federal Regulations. 40 CFR 116:37.

USEPA[a] (1985f) The national pollutant discharge elimination system. 40 CFR 122:101.

USEPA[a] (1985g) Identification and listing of hazardous wastes. 40 CDR 261:384.

USEPA[a] (1986) Guidelines for carcinogenic risk assessment. Fed Regis 51(185):33992–34003. September 24.

Van Duuren BL, Katz C, Goldschmidt BM (1972) Direct-acting alkylating carcinogens—chloro ethers and related compounds. Toxicol Appl Pharmacol 22:279–280.

Van Duuren BL, Goldschmidt BM, Katz C, Seidman I, Paul JS (1974) Carcinogenic activity of alkylating agents. J Natl Cancer Inst 53:695–700.

Van Esch GJ (1981) Induction of preneoplastic lesions in the forestomach of rats after oral administration of 1-chloro-2,3-epoxypropane. I. Range finding studies. Carcinogenicity study. Prepared by Ryksinstitute Voor De Voksgezondheld Bilthoven. Report no. 627805-005.

Van Esch GJ (1982) Induction of preneoplastic lesions in the forestomach of rats after oral administration of 1-chloro-2,3-epoxypropane. II. Carcinogenicity study. Prepared by Ryksinstitute Voor De Voksgezondheld Bilthoven. Report no. 627805-005.

Verschueren K (1983) Handbook of Environmental Data on Organic Chemicals. 2nd Ed. Van Nostrand Reinhold Co., New York, NY, pp 611–613.

Wester PW, Van Der Heiden CA, Bisschop A, Van Esch GJ (1985) Carcinogenicity study with epichlorohydrin (CEP) by gavage in rats. Toxicology 36:325–329.

White AD (1980) *In vitro* induction of sister chromatid exchange in human lymphocytes by epichlorohydrin with and without metabolic activation. Mutat Res 78: 171–176.

Wurgler FE, Graf U (1981) Mutagenic activity of ten coded compounds in the *Drosophila* sex-linked recessive lethal assay. In: de Serres FJ and Ashby J (eds) Evaluation of short-term tests for carcinogens, Elsevier/North Holland Biomedical Press, Amsterdam. pp 666–672.

Control of *Legionella* in Plumbing Systems

I. Introduction

Legionellae are bacteria that have been identified as the cause of legionellosis. Based on a rate of about 1.2 cases of legionellosis/10,000 persons/yr (Foy et al. 1979), more than 25,000 cases of this disease are estimated to occur annually within the U.S. These cases are caused primarily by one of 23 currently recognized species of the genus *Legionella*. Most cases of legionnaires' disease, the pneumonia form of legionellosis, have developed in persons who were immunosuppressed or appeared to be more susceptible because of an underlying illness, heavy smoking, alcoholism, or age (i.e., more than 50 years old). In contrast, although legionnaires' disease has developed in some healthy persons, outbreaks involving healthy persons have been limited to the milder, nonpneumonia form of the disease called Pontiac fever.

Legionellae are widespread in lakes and rivers (Fliermans et al. 1979, 1981), while other studies indicated these organisms may be either sparse or absent in ground water (Fliermans et al. 1982; Spino et al. 1984). Spino et al. (1984) were unable to isolate legionellae after aeration of ground water through a redwood-slat aerator. The possibility that humans may be exposed transiently to legionellae because of a high rate of contact with water is highly probable in view of the high frequency of seropositivity to legionellae in healthy populations (Wentworth et al. 1984) and the widespread occurrence of legionellae in water environments.

In a number of outbreaks of legionellosis in the U.S., aerosols of water documented to contain the specific type of legionellae recovered from the patient were identified as the vehicle for transmission (Cordes et al. 1981; Garbe et al. 1985; Stout et al. 1982). It has been hypothesized that legionellae enter buildings in low numbers via treated drinking water. These bacteria may proliferate in warm water in the presence of factors not yet fully determined. Even when this occurs, as has been shown in numerous buildings, disease usually does not result. Cases and outbreaks of legionellosis occur only when aerosols containing legionellae possessing specific virulence factors (as yet undetermined) are inhaled (possibly ingested) by susceptible persons. Foodborne outbreaks or secondary spread have not been reported.

This Health Advisory discusses the control of legionellae in drinking water,

Reviews of Environmental Contamination and Toxicology, Vol 107

including finished water at treatment facilities, distribution systems, and plumbing systems. Plumbing systems include hot water tanks, taps, showerheads, mixing valves, and faucet aerators, all of which have been associated with the proliferation of legionellae. This guidance does not discuss control of legionellae in whirlpools, respirators, or heat-rejection equipment, such as cooling towers and air conditioners. These, too, have been associated with cases of legionnaires' disease.

II. Presence of Legionellae in Distribution Systems and Plumbing Systems

Legionellae are found in raw water, treated waters, and plumbing systems (Fliermans et al. 1981; Hsu et al. 1984; Witherell et al. 1984), but the occurrence and fate of these organisms in the distribution system between these points are unknown. The organisms may survive a treatment and disinfection process and pass intact through a distribution system. In addition, opportunities exist for the introduction of these organisms into a system by means of broken or corroded piping, repair of existing mainlines, installation of new mains, back siphonage, and cross-connections, any of which may result in contamination of the water supply. In older distribution systems, especially those dependent on gravity flow, deterioration of piping may be so severe that the treated water comes in intimate contact with soil and is subject to infiltration by surface water. Thus, legionellae may be introduced into potable water by any of these routes.

Legionellae surviving initial water treatment may colonize pipe joints and corroded areas or adhere to the surface or sediment of storage tanks, especially those constructed of wood. Here, they may find a habitat suitable for survival and growth (Engelbrecht 1983). Cul-de-sacs, intermittently used storage tanks, and other sites in which waterflow is absent or restricted may also be appropriate habitats for legionellae.

New distribution systems or their components that were not appropriately cleaned and disinfected before use may introduce legionellae into the system. Although this possibility has not been documented, it may not be coincidence that serious outbreaks of legionnaires' disease have occurred in newly opened institutions or buildings (Haley et al. 1979; Helms et al. 1983; Marks et al. 1979). Construction activities may have included intervention in water supply mains with the introduction of contaminated water. The disturbance of sediment and sloughing of scale-bearing high concentrations of legionellae may have occurred by means of hydraulic shock or other perturbations.

Numerous reports of legionellae occurring in plumbing systems, especially hot water systems, have been published. Most of these investigations were carried out in hospitals, and many were prompted by outbreaks of nosocomial (hospital-acquired) legionnaires' disease. In hospitals, legionellae occur primarily in hot water tanks in which water is maintained at temperatures less than 55°C. Legionellae also have been found in showerheads, rubber fittings, aerator screens, faucet spouts, and other plumbing fixtures. These organisms also have been found in residential plumbing systems, such as apartment buildings and homes (Arnow and Weil 1984; Wadowsky 1982), but disease has not been associated with these findings.

III. Control at Water Treatment Facility

Only a few studies have been published on the effectiveness of various types of treatment for eradicating or reducing the numbers of legionellae at a water treatment utility. Tison and Seidler (1983) examined raw water and three types of distribution water supplies: (1) those treated by chlorine (free residual 0.2 to 0.6 mg/L); (2) those treated by sand filtration and chlorination (free residual 0.0 to 0.4 mg/L); and (3) those treated by flocculation, mixed media filtration, and chlorination (free residual 0.5 to 2.0 mg/L). Legionellae organisms were enumerated by direct fluorescent antibody (DFA) tests, and all distribution waters contained about one order of magnitude fewer *Legionella*-like cells than did the raw waters, i.e., 10^3 to 10^4/L. Although evidence suggests that legionellae are common in treated water, the significance of these results is questionable because the authors were unable to isolate any legionellae by animal inoculation or culture procedures, and uncertainties exist regarding specificity of the DFA technique used for legionellae detection.

Most water treatment plants in the U.S. use chlorine disinfection. Although extrapolation of laboratory studies to treatment-plant situations is somewhat tenuous, Kuchta et al. (1983) reported that both *Legionella pneumophila* and *Legionella micdadei* (laboratory-adapted environmental and clinical strains) were much more resistant to chlorine than was *Escherichia coli*. At 21°C, pH 7.6, and 0.1 mg/L of free chlorine residual, a 99% kill was achieved in less than 1 min for *E. coli* compared with 40 min for *L. pneumophila*. Under the same conditions, 0.5 mg/L of free chlorine resulted in a 99.9% legionellae kill in about 5 min. The contact time for a 99% kill of *L. pneumophila* at 4°C was twice as long as as it was at 21°C. The authors concluded that legionellae can survive low levels of chlorine for rather long periods of time.

In a subsequent study, Kuchta et al. (1984) compared agar-passaged (laboratory-adapted) and tap water-grown strains of *L. pneumophila* with respect to chlorine resistance and showed that the latter were considerably more resistant. At 0.25 mg/L free residual chlorine, 21°C, and pH 7.6 to 8.0, a 99% kill of agar-passaged *L. pneumophila* was usually achieved within 10 min, compared with 60 to 90 min for tap water-maintained strains. These data suggest that normal chlorination practices at treatment facilities may not control legionellae.

In contrast to these data, Hsu et al. (1984) reported that survivals of *L. pneumophila* and *E. coli* in various concentrations of chlorine were similar. In an in vitro study, laboratory-adapted strains of *L. pneumophila* Flint 1 serogroup 1 and *E. coli* B were inoculated into several dilutions of sodium hypochlorite in sterile tap water and incubated at 24°C. At 0.2 mg/L residual chlorine, about an order of magnitude reduction occurred in 2 hr for both organisms. Neither organism could be recovered after 2 hr at concentrations equal to or exceeding 2.0 mg/L. The pH values were not reported. The reason for the discrepancy between this study and the Kuchta et al. (1983, 1984) studies may be due to strain or pH differences.

IV. Control of Legionellae in Plumbing Systems

A. Chlorine and Heat

Studies on the control of legionellae in plumbing systems have examined primarily the effectiveness of heat and chlorine. The results of several of these investigations are described here.

In an attempt to eradicate *L. pneumophila* from showers in a transplantation unit experiencing cases of legionnaires' disease, Tobin et al. (1980) emptied the hot and cold water tanks and filled them with water containing 50 mg/L free chlorine. After 3 hr, this process was repeated. Shower fittings were removed and held at 65°C for 18 hr before replacement. Legionellae were not isolated from the shower samples after 6 mon but were found again at 9 mon.

Massanari et al. (1984) controlled a nosocomial outbreak of *L. pneumophila* infection by shock chlorination (15 mg/L) of both hot and cold water supplies for 12 hr. The system was then flushed, and the hot water temperature was raised from 41°C to 64°C for 41 d. These measures significantly reduced the frequency of positive cultures, but 3 of 35 outlets still had positive cultures. Thereafter, a continuous-flow proportional chlorination unit was installed that provided free chlorine levels of 8 and 7.3 mg/L in hot and cold water, respectively. During the first 16 mon of its use, virtually no samples (N = 355) contained *L. pneumophila*, and no new cases of legionellosis were identified. The few positive samples were obtained from rooms that had been vacant for at least 32 d. In this hospital, water was distributed in copper pipes.

Baird et al. (1984) hyperchlorinated a hospital water supply at a constant level of 4 mg/L of free chlorine. The rate of nosocomial legionnaires' disease decreased by almost two-thirds, and the total numbers of legionellae decreased, but the organisms persisted.

Witherell et al. (1984) attempted to eradicate *L. pneumophila* in hospital plumbing by adding chlorine, in proportion to the water demands on the system, to the cold water makeup that supplied the hot water heating system.

This approach was used to avoid corrosion damage resulting from constant-feed chlorination units during periods of low demand. A free chlorine residual of 3.0 mg/L was maintained in the hot water system for 10 d and then reduced to 1.5 mg/L. The organism was not detected by direct culture methods subsequent to disinfection. The corrosivity of the hot water increased slightly (Langelier index = −0.3).

Fisher-Hoch et al. (1981) used hypochlorite to obtain a level of 1 to 2 mg/L of free chlorine at all cold water outlets in a hospital where legionellae were present in both cold and hot water. The free chlorine levels in hot water could not be maintained at more than 0.2 mg/L, and legionellae were recoverable at this level. The water temperature was 45°C, which was warm enough to volatilize the chlorine and cool enough to allow growth of legionellae. Eradication was accomplished successfully by maintaining the hot water temperature at 55 to 60°C, in addition to disinfection of cold water. Subsequently, when a disconnected hot water tank containing stagnant water was turned on again, *L. pneumophila* was found in the water, and a case of nosocomial legionnaires' disease occurred (Fisher-Hoch et al. 1982). A second disconnected tank, which had been drained incompletely, contained a thick, brown liquid deposit at the bottom. This deposit contained 5.4×10^8 *L. pneumophila*/L. Filling the second tank with water containing 50 mg/L of chlorine for 24 hr followed by descaling did not successfully eliminate the legionellae. However, maintaining a constant water temperature of 70°C throughout the tank for 1 hr eliminated the organism. Ciesielski et al. (1984) also noted that legionellae can proliferate in stagnant water inside hot water tanks.

Dennis et al. (1982) examined water samples from the plumbing of 52 hotels, none of which was associated with cases of legionellosis. Ten isolates of *L. pneumophila* were obtained from water samples from eight hotels. Seven of these were from hot water taps or hot-cold mixer showers with water temperatures ranging from 40 to 54°C at the time of sampling. Evidence that these temperatures are not sufficient for *Legionella* control was also provided by Meenhorst et al. (1983). In their study, guinea pigs exposed to aerosolized legionellae from contaminated hot tap water (48°C) contracted pneumonia. The strain of *L. pneumophila* used was isolated fron a series of patients in the Netherlands.

Beam et al. (1984) attempted to control legionellosis outbreaks in two state development centers for the severely handicapped. In one center, hot water tanks that had positive cultures of legionellae were heated to 71°C for 72 hr followed by flushing for 15 min. Because of legionellae regrowth, a monthly heating schedule was established. Subsequently, the chlorine level was raised from 0.5 mg/L to 2 mg/L. This approach was successful in eradicating legionellae from water sources. However, this chlorine level caused leaching from the iron pipes and consequent discoloration of the water. For this reason it was discontinued. Cement liners were installed in the hot water tanks, and the first samples were positive for legionellae. The water temperature was not reported. Soon after, an outbreak of legionellosis occurred.

Plouffe et al. (1983) examined the relationship between the presence of *L. pneumophila* in potable water, nosocomial legionnaires' disease and hot water temperatures in six buildings. *L. pneumophila* was found in the hot water of all four buildings in which hot water was maintained at 43 to 49°C (110 to 120°F), and nosocomial legionnaires' disease was found in three of these buildings. No organism and no disease was found in the two buildings where hot water was maintained at 57 to 60°C (135 to 140°F). The plumbing system of one of the buildings with both *L. pneumophila* and legionnaires' disease was flushed with 71°C water, and the hot water was thereafter maintained at 57 to 60°C. No new cases of legionnaires' disease occurred for at least 6 mon. The authors concluded that colonization and nosocomial legionnaires' disease can be prevented by maintaining hot water at 57 to 60°C.

In another attempt to eradicate *L. pneumophila* and nosocomial legionnaires' disease, Yu et al. (1982) raised the temperature in hot water storage tanks from 45 to 60°C for 72 hr and flushed 50 showers and 360 faucets for 20 min with 60°C water to eliminate the organisms from the sediment. Colony counts were substantially reduced. After 3 mon, colony counts increased rapidly from 4 colonies/mL to more than 300 colonies/mL, and nosocomial legionnaires' disease again appeared. The authors concluded that a periodic schedule of short-term temperature elevation of the hot water system may control nosocomial legionnaires' disease.

Stout et al. (1986) tested 75 legionellae isolates for their ability to withstand high temperatures. Tubes containing buffered yeast extract broth, sterile water, or hot-water-tank water plus sediment were inoculated and placed in 60°C, 70°C, or 80°C water baths. At 60°C, 4 min were required for a 1-log reduction of *L. pneumophila* in the water plus sediment tube. Approximately 25 min were required at this temperature to sterilize a suspension of *L. pneumophila* containing 10^8 colonies/mL. The authors recommended that a flush temperature exceeding 60°C be maintained for at least 30 min when flushing distal outlets.

Muraca et al. (1987) compared the relative efficacies of heat (60°C), ozone (1 to 2 mg/L), ultraviolet (UV) light (30,000 μW-sec/cm^2 at 254 nm) and hyperchlorination (4 to 6 mg/L) to eradicate *L. pneumophila* in a model plumbing system. Nonturbid water at 25°C and 43°C and turbid water at 25°C were tested. When samples were taken of the circulated water, a 5-log kill of a 10^7 bacteria/mL concentration was achieved with all treatments within 6 hr. However, it is noteworthy that heat completely eradicated *Legionella* in less than 3 hr; whereas UV light produced a 5-log decrease in 20 min and no further inactivation was seen during the 6-hr observation period. Chlorine and ozone required 5 hr to effect a similar 5-log decrease, and chlorine achieved complete eradication in only the nonturbid samples during the 6 hr. Ozone killed the organisms in both turbid and nonturbid water in 4 to 5 hr.

B. Ozone Treatment

Edelstein et al. (1982) used ozone in an attempt to eradicate legionellae from the potable water supply of an unused wing of a hospital that was known to be contaminated with bacteria. The results were inconclusive because the organisms were eliminated from both the experimental wing and the control wing that was untreated. The latter was believed to be due to excess mechanical flushing and an unexpected rise in the chlorine content of the main water supply. The in vitro susceptibility of *L. pneumophila* to ozone was about 0.36 mg ozone/L, but was not consistent. The ozone mean residual concentration used in the hospital water system was 0.79 mg/L.

C. Ultraviolet Radiation Treatment

Antopol and Ellner (1979) reported that 90% of *L. pneumophila* cells in distilled water were killed by 920 μW-sec/cm^2 UV radiation. This could be compared with exposures ranging from 2,100 to 5,000 μW-sec/cm^2 for killing of *E. coli*, and *Salmonella*, *Serratia*, and *Pseudomonas* spp. The latter values obtained under the same conditions as those used for *L. pneumophila* indicate that legionellae may be more than twice as susceptible to UV radiation than are the other organisms.

Gilpin (1984) reported laboratory and field experiments using UV radiation to inactivate *Legionella* spp. in standing and recirculating water systems. Times of exposure to 1 μW/cm^2 of UV radiation to produce 90% killing of six species of *Legionella* ranged from 17 to 44 min. A commercial UV apparatus killed 99% of the organisms in less than 30 sec in a 3-L recirculating water system.

In addition, Knudson (1985) reported that when agar plates seeded with *L. pneumophila* were exposed to 240 μW/cm^2 for 25 sec or less, a reduction of 6 to 7 orders of magnitude was observed. However, when UV-irradiated legionellae were exposed to indirect sunlight for 60 min, the recovery rates were 2 orders of magnitude greater than those not exposed to sunlight, as a result of photoreactivation.

D. Ethylene Oxide Treatment

Cordes et al. (1981) sterilized *Legionella*-contaminated showerheads with ethylene oxide, but they were soon recontaminated.

V. Design of Hot Water Tanks

Often legionellae have been reported in hot water tanks, particularly in the bottom sediment. The design of these tanks is important in the control of these bacteria. Most residential hot water tanks are heated from the bottom near the cold water entrance pipe and are more likely to maintain a bottom temperature high enough (> 55°C) to prevent growth of legionellae. However, when thermostats in homes are set low (< 55°C) as an energy conservation measure, growth of legionellae may result.

Thermostats for hot water heaters in hospitals and other health-care facilities are usually set at lower temperatures to comply with the recommendation of the Joint Commission on Accreditation of Hospitals that the water temperature be "safe" (JCAH 1985). This practice, which is performed to prevent scalding of patients using hot water, may promote growth of legionellae. Large institutional tanks also are heated more often by internal steam coils or by other heaters located midway from top to bottom of the tank. The water at the bottom may not be heated sufficiently to kill legionellae. Periodic partial draining of these tanks from the bottom to eliminate sediment may control proliferation of legionellae. This is especially important, because environmental microflora in the sediment are known to produce metabolites, possibly including cysteine, that stimulate growth of legionellae (Stout et al. 1985). Removal from other areas of the plumbing system where water stagnates may also prevent or control growth of legionellae (Stout et al. 1985).

VI. Type of Water Fittings

Information on the specific types of gaskets and fittings that support the colonization of legionellae is not well documented. One study of water fittings as sources of *L. pneumophila* in a hospital plumbing system was carried out by Colburne et al. (1984a,b). In well-controlled experiments, *L. pneumophila* was isolated from rubber washers and gaskets but not from fiber or plastic fittings. The ability of the bacteria to multiply when in contact with the rubber fittings was demonstrated. When the rubber fittings were replaced with plastic fittings, *L. pneumophila* could not be isolated up to 1 yr later. The authors concluded that shower and tap fittings that support growth of legionellae provide habitats protected from chlorine and heat. These foci may be seeded constantly or intermittently with legionellae from hot water tanks or other amplifiers within the distribution system.

VII. When to Control Legionellae in Plumbing Systems

Legionellae are often found in the plumbing systems of hospitals that have not experienced any cases of legionnaires' disease. One reason may be that some strains are more virulent than others. No practical method is now available to distinguish the virulent strains from avirulent strains. For this reason, some experts believe that the mere presence of legionellae in the absence of disease is not sufficient grounds to undertake control measures (Jakubowski et al. 1984). They believe that health-care institutions should focus initially on surveillance for respiratory illness, especially in high-risk patients, rather than control legionellae in plumbing systems. If nosocomial legionellosis is identified and environmental strains match patient isolates, control in plumbing systems is indicated.

In contrast, Edelstein (1985) stated that most authorities would probably agree that disinfection of a contaminated site is indicated when: (1) the site is implicated as a source of an outbreak of legionnaires' disease or Pontiac fever; (2) contamination is present in a hospital ward, especially one housing high-risk patients, such as an organ transplantation unit, regardless of epidemiologic findings (selective decontamination of certain ward areas may be feasible); and (3) contamination is found in a building that has not been used for some time and in which the water has stagnated.

Because of the virulence of some of these strains and the annual occurrence of at least 25,000 cases or more in the U.S., a stronger preventive approach could be supported.

No consensus exists as to when measures should be undertaken to control legionellae in the plumbing system of health-care institutions. Once virulence factors can be identified and virulent strains differentiated from avirulent strains, routine monitoring of the plumbing system may become more practical. Until then, the USEPA Office of Drinking Water recommends that, on the basis of the high incidence and mortality rate, health care institutions consider preventive measures for the control of legionellae in their plumbing systems. These measures could also control other opportunistic pathogens in the system that might cause nosocomial infections.

VIII. Summary

Legionellae are abundant in ambient water and may survive water treatment, especially because they are relatively resistant to chlorine. Once in treated water, they pass, probably at low levels, through the distribution system. It is also possible that legionellae enter the distribution system through broken or corroded piping, repair of existing mains, installation of new mains, back siphonage, and cross-connections. When legionellae enter hot water tanks, they settle to the bottom; under certain circumstances, they proliferate. If they proliferate, plumbing fixtures, such as aerators, water fittings, and showerheads, may be seeded, resulting in colonization and growth at these sites.

Inhalation of aerosolized potable water was suggested, based on data from outbreak investigations, as a primary route of infection, although ingestion is also a possibility. The most susceptible individuals are those with underlying diseases, especially those involving immunosuppressive therapy. In several outbreaks, however, legionellosis developed in apparently healthy persons. Other risk factors include alcohol abuse, surgery, and smoking.

To reduce legionellae levels in drinking water, the presence of organic matter and growth of algae and protozoa should be minimized in storage reservoirs. Moreover, newly repaired or constructed components of a water distribution system should be flushed thoroughly and disinfected before operation begins. Even after flushing and disinfection, it cannot be assumed that legionellae have been controlled, because design factors in the distribution system may impede efficiency of these measures.

To control legionellae growth in hot water plumbing, several approaches may be considered. Most of the published data have examined the effectiveness of chlorine or heat or both. The maintenance of free chlorine has been found effective for controlling legionellae. Shock chlorination also is effective, but unless free chlorine is maintained within a system, the organism may reappear. Control probably can be achieved if free chlorine levels in hot water are maintained at 8 mg/L, but at this level corrosion of pipes may occur. In some cases, control may be achieved at 1.5 to 2 mg/L free chlorine. Undoubtedly, the level of chlorine found effective depends, in part, on the design criteria of the plumbing system. A pertinent facet in controlling legionellae is the difficulty of controlling batch chlorination and of maintaining a chlorine residual in hot water. This problem can be minimized by using a continuous-flow proportional chlorinator in hot water systems.

Heat shock may eradicate legionellae in hot water tanks if the temperature at the bottom of the tank is maintained at 70°C for 1 hr, but this is a temporary measure that must be performed routinely to be effective. Maintenance of hot water at 55°C or higher apparently controls the organism, although lower temperatures may not. If legionellae are controlled by heat, care must be taken to prevent scalding of persons using the water, especially in health care institutions.

Disinfection of a plumbing system by heat treatment or chlorine treatment alone may not be as effective as a combination of the two. For example, growth of legionellae may theoretically be enhanced on the cold water side of a hot-cold water mixing valve in a heat-treated plumbing system, and chlorine may be effective in this location.

Effective disinfection of legionellae by ozone, UV radiation, or ethylene oxide has not been demonstrated by field tests.

In addition to chemical and heat disinfection, other procedures may be effective in controlling legionellae. Hot water tanks should be designed to provide uniform temperatures throughout. Hot or cold water tanks used intermittently should be disconnected from the system, drained, flushed, and disinfected before being reconnected. Hot water tanks should be drained regularly or at least bled to remove accumulated sludge that may serve as a substrate for growth of legionellae and other microorganisms. Taps and showers in unused areas of health care facilities should at least be flushed before patients are exposed to them. Finally, faucet sieves and aerators, and rubber washers and gaskets in the plumbing system, should be used with caution, especially in institutions housing physically compromised individuals and in which hot water is maintained at temperatures lower than 55°C.

References

Antopol SC, Ellner PD (1979) Susceptibility of *Legionella pneumophila* to ultraviolet radiation. Appl Environ Microbiol 38:347–348.

Arnow PM, Weil D (1984) *Legionella pneumophila* contamination of residential tap water. In: Thornsberry C, Balows A, Feeley JC, Jakubowski W (eds) Legionella: Proceedings of the 2nd international symposium, June 19–23, 1983, Atlanta, GA. American Society for Microbiology, Washington, DC, pp 240–241.

Baird IM, Potts W, Smiley J, Click N, Schleich S, Connole C, Davidson K (1984) Control of endemic nosocomial legionellosis by hyperchlorination of potable water. In: Thornsberry C, Balows A, Feeley JC, Jakubowski W (eds) Legionella: Proceedings of the 2nd international symposium, June 19–23, 1983, Atlanta, GA. American Society for Microbiology, Washington, DC, p 333.

Beam TR, Moreton D, Raab TA, Heaslip W, Montes M, Hanrahan J, Best M, Yu VL (1984) Epidemiology and control of *Legionellaceae* in state developmental centers. In: Thornsberry C, Balows A, Feeley JC, Jakubowski W (eds) Legionella: Proceedings of the 2nd international symposium, June 19–23, 1983, Atlanta, GA. American Society for Microbiology, Washington, DC, pp 236–237.

Ciesielski CA, Blaser MJ, Wang W-LL (1984) Role of stagnation and obstruction of water flow in isolation of *Legionella pneumophila* from hospital plumbing. Appl Environ Microbiol 48:984–987.

Colbourne JS, Smith MG, Fisher-Hoch SP, Harper D (1984a) Source of *Legionella pneumophila* infection in a hospital hot water system: materials used in water fittings capable of supporting *L. pneumophila* growth. In: Thornsberry C, Balows A, Feeley JC, Jakubowski W (eds) Legionella: Proceedings of the 2nd international symposium, June 19–23, 1983, Atlanta, Georgia. American Society for Microbiology, Washington, DC, pp 305–307.

Colbourne JS, Pratt DJ, Smith MG, Fisher-Hoch SP, Harper D (1984b) Water fittings as sources of *Legionella pneumophila* in a hospital plumbing system. Lancet 1:210–213.

Cordes LG, Wiesenthal AM, Gorman GW, Phair JP, Sommers HM, Brown A, Yu VL, Magnussen MH, Meyer RD, Wolf JS, Shands KN, Fraser DW (1981) Isolation of *Legionella pneumophila* from hospital shower heads. Ann Intern Med 94(2):195–197.

Dennis PJ, Taylor JA, Fitzgeorge RB, Bartlett CLR, Barrow GI (1982) *Legionella pneumophila* in water plumbing systems. Lancet 1:949–951.

Edelstein PH (1985) Environmental aspects of *Legionella* ASM News 51:460–467.

Edelstein PH, Whittaker RE, Kreiling RL, Howell CL (1982) Efficacy of ozone in eradication of *Legionella pneumophila* from hospital plumbing fixtures. Appl

Environ Microbiol 44:1330–1334.

Engelbrecht RS[a] (1983) Source, treatment, and distribution. In: Berger PS, Argaman Y (eds) Assessment of microbiology and turbidity standards for drinking water. USEPA Report no. EPA-570/9-83-001a. U.S. Environmental Protection Agency, pp 1–68.

Fisher-Hoch SP, Bartlett CLR, Tobin J O'H, Gillett MB, Nelson AM, Pritchard JE, Smith MG, Swann RA, Talbot JM, Thomas JA (1981) Investigation and control of an outbreak of legionnaires' disease in a district general hospital. Lancet 1: 932–936.

Fisher-Hoch SP, Smith MG, Colbourne JS (1982) *Legionella pneumophila* in hospital hot water cylinders [letter]. Lancet 1:1073.

Fliermans CB, Cherry WB, Orrison LH, Thacker L (1979) Isolation of *Legionella pneumophila* from nonepidemic-related aquatic habitats. Appl Environ Microbiol 37:1239–1242.

Fliermans CB, Cherry WB, Orrison LH, Smith SJ, Tison DL, Pope DH (1981) Ecological distribution of *Legionella pneumophila*. Appl Environ Microbiol 41:9–16.

Fliermans CB, Bettinger GE, Fynsk AW (1982) Treatment of cooling systems containing high levels of *Legionella pneumophila*. Water Res 16:903–909.

Foy HM, Hayes PS, Cooney MK, Broome CV, Allan I, Robe R (1979) Legionnaires' disease in a prepaid medical-care group in Seattle, 1963–75. Lancet 1:767–770.

Garbe PL, Davis BJ, Weisfeld JS, Markowitz L, Miner P, Garrity F, Barbaree JM, Reingold AL (1985) Nosocomial legionnaires' disease: epidemiologic demonstration of cooling towers as a source. JAMA 254:521–524.

Gilpin RW (1984) Laboratory and field applications of ultra-violet light disinfection on six species of *Legionella* and other bacteria in water. In: Thornsberry C, Balows A, Feeley JC, Jakubowski W (eds) Legionella: Proceedings of the 2nd international symposium, June 19–23, 1983, Atlanta, GA. American Society for Microbiology, Washington, DC, pp 337–339.

Haley CE, Cohen ML, Halter J, Meyer RD (1979) Nosocomial legionnaires' disease: a continuing common-source epidemic at Wadsworth Medical Center. Ann Intern Med 90:583–586.

Helms CM, Massanari RM, Zeitler R, Streed S, Gilchrist MJR, Hall N, Hausler WJ, Sywassink J, Johnson W, Wintermeyer L, Hierholzer WJ (1983) Legionnaires' disease associated with a hospital water system: a cluster of 24 nosocomial cases. Ann Intern Med 99(2):172–178.

Hsu SC, Martin R, Wentworth BB (1984) Isolation of *Legionella* species from drinking water. Appl Environ Microbiol 48:830–832.

Jakubowski W, Broome CV, Geldreich EE, Dufour AP (1984) Transmission and control. In: Thornsberry C, Balows A, Feeley JC, Jakubowski W (eds) Legionella: Proceedings of the 2nd international symposium, June 19–23, 1983, Atlanta, GA. American Society for Microbiology, Washington, DC, pp 351–355.

JCAH (1985) Joint Commission on Accreditation of Hospitals. Accreditation manual for hospitals, chapter on plant, technology, and safety management JCAH, Chicago, IL.

Knudson GB (1985) Photoreactivation of UV-irradiated *Legionella pneumophila* and other *Legionella* species. Appl Environ Microbiol 49:975–980

Kuchta JM, States SJ, McGlaughlin JE, Wadowsky RM, McNamara AM, Wolford

RS, Yee RB (1984) Enhanced chlorine resistance of tapwater grown *Legionella pneumophila* as compared with agar-passaged strains. Paper no. Q2. Abstracts, annual meeting of the American Society for Microbiology, St. Louis, MO.

Kuchta JM, States SJ, McNamara AM, Wadowsky RM, Yee RB (1983) Susceptibility of *Legionella pneumophila* to chlorine in tapwater. Appl Environ Microbiol 46: 1134–1139.

Marks JS, Tsai TF, Martone WJ, Baron RC, Kennicott J, Holtzhauer FJ, Baird I, Fay D, Feeley JC, Mallison GF, Fraser DW, Halpin TJ (1979) Nosocomial legionnaires' disease in Columbus, OH. Ann Intern Med 90(4):565–569.

Massanari RM, Helms C, Zeitler R, Streed S, Gilchrist M, Hall N, Hausler W, Johnson W, Wintermeyer L, Muhs JS, Hierholzer WJ (1984) Continuous hyperchlorination of potable water systems for control of nosocomial *Legionella pneumophila* infections. In: Thornsberry C, Balows A, Feeley JC, Jakubowski W (eds) Legionella: Proceedings of the 2nd international symposium, June 19–23, 1983, Atlanta, GA. American Society for Microbiology, Washington DC, pp 334–336.

Meenhorst PL, Reingold AL, Gorman GW, Feeley JC, van Cronenburg BJ, Meyer CLM, van Furth R (1983) *Legionella* pneumonia in guinea pigs exposed to aerosols of concentrated potable water from a hospital with nosocomial legionnaires' disease. J Infect Dis 147(1):129–132.

Muraca P, Stout JE, Yu VL (1987) Comparative assessment of chlorine, heat, ozone, and UV light for killing *Legionella pneumophila* within a model plumbing system. Appl Environ Microbiol 53:447–453.

Plouffe JF, Webster LR, Hackman B, Macynski M (1983) Hot water temperature, *L pneumophila* (LP) and nosocomial legionnaires' disease (LD). Paper L16. Abstracts, annual meeting of the American Society for Microbiology, New Orleans, LA.

Spino DF, Rice EW, Geldreich EE (1984) Occurrence of *Legionella* spp. and other aquatic bacteria in chemically contaminated ground water treated by aeration. In. Thornsberry C, Balows A, Feeley JC, Jakubowski W (eds) Legionella: Proceedings of the 2nd international symposium, June 19–23, 1983, Atlanta, GA. American Society for Microbiology, Washington, DC, pp 318–320.

Stout JE, Best MG, Yu VL (1986) Susceptibility of members of the family *Legionellaceae* to thermal stress: implications for heat eradication methods in water distribution systems. Appl Environ Microbiol 52.396–399.

Stout J, Yu VL, Vickers RM, Zuravleff J, Best M, Brown A, Yee RB, Wadowsky R (1982) Ubiquitousness of *Legionella pneumophila* in the water supply of a hospital with endemic Legionnaires disease. N Engl J Med 306:466–468.

Stout JE, Yu VL, Best MG (1985) Ecology of *Legionella pneumophila* within water distribution systems. Appl Environ Microbiol 49·221–228

Tison DL and Seidler RJ (1983) *Legionella* incidence and density in potable drinking water supplies. Appl Environ Microbiol 45:337–339.

Tobin J O'H, Beare J, Dunnill MS, Fisher-Hoch S, French M, Mitchell RG, Morris PJ, Muers MF (1980) Legionnaires' disease in a transplant unit isolation of the causative agent from shower baths. Lancet 2.118–121

Wentworth BB, Chadwick WA, Stiefel HE, Benge DS (1984) Prevalence of antibody to various *Legionella* species in ill and healthy populations in Michigan In: Thornsberry C, Balows A, Feeley JC, Jakubowski W (eds) Legionella: Proceedings of the

2nd international symposium, June 19–23, 1983, Atlanta, GA. American Society for Microbiology, Washington, DC. pp 255–257.

Wadowsky RM, Yee RB, Mezmar L, Wing EJ, Dowling JN (1982) Hot water systems as sources of *Legionella pneumophila* in hospital and nonhospital plumbing fixtures. Appl Environ Microbiol 43: 1104–1110.

Witherell LE, Orciari LA, Duncan RW, Stone KM, Lawson JM (1984) Disinfection of hospital hot water systems containing *Legionella pneumophila*. In: Thornsberry C, Balows A, Feeley JC, Jakubowski W (eds) Legionella: Proceedings of the 2nd international symposium, June 19–23, 1983, Atlanta, GA. American Society for Microbiology, Washington, DC, pp 336–337.

Yu VL, Best M, Stout J, Brown A, Goetz A (1982) Effectiveness of intermittent short-term temperature elevation of the hospital water supply in controlling nosocomial legionnaires' disease (LD). Paper L19. Abstracts, annual meeting of the American Society for Microbiology, Atlanta, GA.

Mercury[a]

I. General Information

Two of the most important ionic mercury compounds are mercury (II) chloride, also known as mercuric bichloride and mercury perchloride, and mercury (II) sulfate, which is sometimes called mercuric sulfate. This advisory is primarily concerned with the toxic effects of such ionic compounds, but it is metallic mercury that has the most uses. Metallic mercury is used as a cathode in the electrolytic preparation of chlorine and caustic soda; it is also used in electrical apparatuses, dental amalgams, and catalysts, and in pulp and paper manufacture. The properties of mercury, mercury (II) chloride, and mercury (II) sulfate are given in Tables 1, 2, and 3, respectively.

Mercury, although a relatively rare element, is ubiquitous in the earth's crust, occurring at levels from 10 to 500 ppb as a sulfide, chloride, or oxide. Mercury can form organic compounds that bioaccumulate in the food chain and become a significant toxicological concern. Only a small fraction of mercury in ground and surface waters occurs in the organic form (USEPA 1987).

Most mercury used commercially in the United States is imported. Commercial uses have resulted in releases of mercury and its compounds to surface waters. Naturally occurring levels of mercury in ground and surface water are less than 0.5 μg/L, although higher levels may occur in ground water from local mineral deposits. Ground-water surveys have found mercury at levels above 0.5 μg/L in 15 to 30% of wells tested. Surface water surveys report that about 20% of surface waters exceed 0.5 μg/L. State compliance data show that 16 ground-water and 16 surface-water wells currently exceed the maximum contaminant level of 2 μg/L (USEPA 1987).

II. Pharmacokinetics

A. Absorption

An estimated 7 to 15% of orally administered inorganic mercury is absorbed by humans (Rahola et al. 1971; Task Group on Metal Accumulation 1973).

[a] Based on USEPA (1985), except for Section I which is based on USEPA (1987)

Reviews of Environmental Contamination and Toxicology, Vol. 107

Table 1. General information and properties of mercury

Property	Value
Chemical Abstracts Service (CAS) number	7439-97-6
Chemical formula	Hg
Molecular weight	200.59
Physical state	Silver liquid
Boiling point	356.58°C
Melting point	−38.87°C
Density	13.59
Vapor pressure	2×10^{-3} mm (25°C)
Specific gravity	13.59
Water solubility	0.35 mg/L (25°C)[a]
Log octanol/water partition coefficient	—
Odor threshold	—
Taste threshold	—

Source Weast (1971)
[a] Seiler (1988)

Table 2. General information and properties of mercury (II) chloride[a]

Property	Value
Chemical Abstracts Service (CAS) number	7487-94-7
Chemical formula	$HgCl_2$
Molecular weight	271.49
Physical state	White powder
Boiling point	302°C
Melting point	276°C
Density	5.44
Vapor pressure	1 mm (136.2°C)
Specific gravity	5.44
Water solubility	6.9 g/100 mL (20°C)
Log octanol/water partition coefficient	—
Odor threshold	—
Taste threshold	—

Source Weast (1971)

Table 3. General information and properties of mercury (II) sulfate[a]

Property	Value
Chemical Abstracts Service (CAS) number	7783-36-0
Chemical formula	$HgSO_4$
Molecular weight	296.65
Physical state	White powder
Boiling point	—
Melting point	Decomposes
Density	6.47
Vapor pressure	—
Specific gravity	6.47
Water solubility	Decomposes
Log octanol/water partition coefficient	—
Odor threshold	—
Taste threshold	—

Source· Weast (1971).

B. Distribution

Rothstein and Hayes (1960) administered a single iv dose of ^{203}Hg [as $Hg(NO_3)_2$; 0.2 mg/kg] to seven male Wistar rats. Mercury was distributed primarily to kidney, liver, blood, skin, and muscle. Other tissues contained only fractional percentages of the administered dose. In general, each tissue except the kidney showed a maximum value 4 hr or 1 d post-treatment, followed by rapid clearance. The kidney continued to accumulate mercury, with maximum concentrations reached at 6 to 15 d. For example, after 4 hr, only 9% of the body burden was found in the kidney; by d 15 post-treatment 86% of the remainder was in the kidney.

Jugo (1976) administered single iv injections of ^{203}Hg (as $HgCl_2$; 0.15 mg/kg) to 2- or 21-wk-old female albino rats (strain not specified). Approximately 28 and 51% of the administered dose was present in the kidneys of the 2- and 21-wk-old rats, respectively, after 144 hr. Approximately 9% of the dose was present in the liver of 2-wk-old rats; less than 1% was present in the liver of older rats. In both groups of rats, blood and brain contained less than 1% of the administered dose.

C. Metabolism

1. Metabolism. No information is available on the biotransformation of inorganic mercury in the literature. However, it is known that nearly all ionic mercury entering the body binds to the thiol groups in proteins and other biological molecules (Clarkson 1972).

D. Excretion

Rahola et al. (1971) administered single oral doses of inorganic mercury (6 μg/subject) and protein-bound methyl mercury (14 μg/subject) to human volunteers. Approximately 85% of the inorganic mercury was eliminated in the feces within 4 to 5 d; only about 0.2% was excreted in the urine. After 50 d the daily excretion of inorganic mercury in the urine and feces was about 0.02% of the dose. Approximately 6% of the dose of methyl mercury was eliminated in the feces within 3 to 4 d; excretion in the urine was negligible at first but increased slowly. After 100 d, 20% of the daily excretion of mercury was via the urine.

Rothstein and Hayes (1960) studied the excretion of mercury in rats administered single iv injections of ^{203}Hg [as $Hg(NO_3)_2$; 0.2 mg/kg]. The clearance of mercury occurred in three phases: a rapid phase lasting a few days and eliminating 35% of the dose; a slower phase with a half-life of 30 d eliminating 50% of the dose; and a slow phase (half-life of approximately 100 d) eliminating the remaining 15%. Since mercury accumulated in the kidney in the first few days following dosing, the two slow phases of excretion primarily represent clearance from the kidney.

III. Health Effects

A. Humans

1. Short-Term Exposure. Gleason et al. (1957) estimated that the lethal oral dose for mercuric salts in humans is 1 to 4 g (equivalent to 14 to 57 mg/kg).

Ingestion of a dose of 1.5 g of mercuric chloride ($HgCl_2$) produced emesis after 5 min, followed by severe abdominal pain and a brief period of loss of consciousness (Pesce et al. 1977).

2. Long-Term Exposure. No information was located in the literature searched on the health effects of long-term exposure to inorganic mercury.

B. Animals

1. Short-Term Exposure. Male and female brown Norway rats (varying numbers per dose group) were given subcutaneous injections of mercuric chloride, 3 times/week for a maximum of 12 wk. The dose levels administered were 0, 0.05, 0.1, 0.25, 0.5, 1.0, or 2.0 mg/kg/injection. Rats that received doses of 0.1 mg/kg/injection or higher developed renal disease characterized by antiglomerular basement membrane antibodies and deposits in the glomerular tufts and small renal arteries. Proteinuria and a nephrotic syndrome were also observed. Based on these results, a No-Observed-Adverse-Effect Level (NOAEL) of 0.05 mg/kg/injection was identified (Druet et al. 1978).

Bernaudin et al. (1981) reported that mercurials administered by ingestion or inhalation to Brown Norway rats resulted in a systemic auto-immune disease. Rats were fed 0 or 3 mg/kg/wk of mercuric chloride for 60 d. Immunofluorescence histology showed that four of five treated animals had a linear IgG deposition in the glomeruli after 15 d of exposure. After 60 d of exposure, all the animals were seen with mixed or linear deposition of IgG in the glomeruli and granular IgG in the arteries. Weak proteinuria was observed in three of five rats. The results of this study indicate a LOAEL of 3 mg/kg/wk.

2. Long-Term Exposure. Male and female rats (strain not specified; 20 to 24 per group) were administered mercury as mercuric acetate in the diet for up to 2 yr at concentrations of 0, 0.5, 2.5, 10, 40, or 160 ppm. Assuming that 1 ppm in the diet of rats is equivalent to 0.05 mg/kg/d (Lehman 1959), these dose levels correspond to 0, 0.025, 0.125, 0.50, 2.0, or 8.0 mg/kg/d. At the highest dose, body weight was slightly depressed in male rats only (statistical significance not specified). Kidney weights were significantly increased ($p < 0.05$) in the 2.0- and 8.0-mg/kg/d dose. Pathological changes originating in the proximal convoluted tubules were also noted at these dose levels, with more severe effects in females than in males. Based on these results, a NOAEL of 10 ppm (0.5 mg/kg/d) was identified. A number of deficiencies limit the usefulness of this study: only a small number of animals survived past 18 mon, information on the number of animals in each group that had detectable pathological changes is missing, and body weight changes in males were not statistically analyzed (Fitzhugh et al. 1950).

Andres (1984) administered mercuric chloride (3 mg/kg bw) by gavage (in water) to five Brown Norway rats and two Lewis rats twice a wk for 60 wk. Two Brown Norway rats died after 30 to 40 d. Standard histopathology of all the animals was normal after 60 d of exposure. Immunofluorescence, however, revealed deposits of IgG in the renal glomeruli of the mercuric chloride-treated Brown Norway rats. The Brown Norway rats also exhibited mercury-induced morphological lesions of the ileum and colon with abnormal deposits of IgA in the basement membranes of the intestinal glands. No adverse effects were noted in the Lewis rats. Based on the observations from this study, a LOAEL of 3 mg/kg is identified for Brown Norway rats.

3. Reproductive Effects. No information was located in the literature searched on the reproductive effects of inorganic mercury.

4. Developmental Effects. Oral dosing of Syrian golden hamsters with mercuric acetate on d 8 of gestation at levels ranging from 4 to 100 mg/kg produced a dose-increase response in numbers of resorptions and abnormal embryos. While these findings were evident at the 4-mg/kg dose level, the percentage of change was not significantly different from that in controls at this low level (Gale 1974).

5. *Mutagenicity.* No evidence is available to indicate that the mercuric salts are mutagenic.

6. *Carcinogenicity.* No evidence is available on the carcinogenicity of inorganic mercury.

IV. Quantification of Toxicological Effects[a]

A. One-Day Health Advisory

The available data are insufficient to develop a One-Day Health Advisory (HA) for mercury.

B. Ten-Day Health Advisory

The available data are insufficient to develop a Ten-Day HA for mercury.

C. Longer-Term Health Advisory

The available data are insufficient to develop Longer-Term HAs for mercury.

D. Lifetime Health Advisory

Considering the results of the Druet et al. (1978), Bernaudin et al. (1981), and Andres (1984) studies, a LOAEL of 3 mg/kg was used as the basis for calculating the Lifetime HA. Using a factor of 0.739 to adjust the exposure for mercuric ion from mercuric chloride in a mg/kg/d dose, the RfD may be calculated as follows:

Step 1. Determination of the Reference Dose (RfD)

$$\text{RfD} = \frac{(3\ \text{mg/kg})(0.739)}{(7\ \text{d})(1{,}000)} = 0.0003\ \text{mg/kg/d}$$

Step 2. Determination of the Drinking Water Equivalent Level (DWEL)

$$\text{DWEL} = \frac{(0.0003\ \text{mg/kg/d})(70\ \text{kg})}{(2\ \text{L/d})} = 0.011\ \text{mg/L}\ (10\ \mu\text{g/L})$$

Step 3: Determination of the Lifetime Health Advisory

$$\text{Lifetime HA} = (0.01\ \text{mg/L})(20\%) = 0.002\ \text{mg/L}\ (2\ \mu\text{g/L})$$

[a] See Introduction for discussion of methodology

E. Evaluation of Carcinogenic Potential

The International Agency for Research on Cancer has not evaluated the carcinogenic potential of mercury. Using the criteria described in USEPA's guidelines for assessment of carcinogenic risk (USEPA 1986), mercury may be classified in Group D: not classified. This group is for substances with inadequate animal evidence of carcinogenicity.

V. Other Criteria, Guidance, and Standards

The USEPA has recommended an ambient water quality criterion for the protection of health of 144 ng/L (USEPA 1980) and for drinking water of 2 μg/L (USEPA 1973). A World Health Organization expert group has recommended an international standard for mercury in drinking water at 1 μg/L (WHO 1971).

VI. Analytical Methods

Mercury levels are determined by flameless atomic absorption using either a manual cold vapor technique (USEPA 1979a) or an automated cold vapor technique (USEPA 1979b). The flameless atomic absorption procedure is a physical method based on the absorption of radiation at 253.7 nm by mercury vapor. The mercury is reduced to the elemental state and aerated from solution in a closed system. The mercury vapor passes through a cell positioned in the light path of an atomic absorption spectrophotometer. Absorbance is measured as a function of mercury concentration. The detection limit for mercury is 0.2 μg/L using either the manual or automated technique.

VII. Treatment Technologies

Laboratory and pilot plant studies indicate that coagulation/filtration is moderately effective in the removal of inorganic mercury from drinking water. Ferric sulfate coagulation achieved 66% removal of pH 7 and 97% removal at pH 8 from water containing 0.05 mg/L of inorganic mercury. Alum coagulation was much less effective: 47% of the mercury was removed at pH 7 and 38% at pH 8. Coagulation/filtration is less effective, however, for the removal of organic mercury, although the efficiency of existing systems can be improved by the addition of powdered-activated carbon (PAC) to the raw water influent. Laboratory tests by Logsdon and Symons (1973) have shown that each milligram per liter of PAC removes 0.0001 mg/L of either inorganic or organic mercury.

Lime softening is essentially ineffective for the removal of organic mercury, but it is moderately effective for the removal of inorganic mercury, depending on the pH of the water. Laboratory studies by Logsdon and Symons (1973) have shown that in the 10.7 to 11.4 pH range, lime softening removed 60 to 80% of the inorganic mercury, whereas only about 30% removal was achieved at pH 9.4.

The use of activated carbon to remove mercury from drinking water has been reported by various investigators (Logsdon and Symons 1973; Sigworth and Smith 1972; Sorg 1979; Theim et al. 1976). Laboratory tests were performed by pumping solutions of tap water and either soluble inorganic or organic mercury through columns of granular-activated carbon (GAC) for extended periods of time. The results showed that 80 to 99% of the mercury is removed from the water by this technology (Logsdon and Symons 1973; Sigworth and Smith 1972).

Limited pilot-plant studies have been reported by Sorg (1977) on the use of reverse osmosis for mercury removal. One study investigating the removal of heavy metals, pesticides, and other toxic chemicals from secondary wastewater effluent resulted in inorganic and organic mercury removals of 82 and 83%, respectively. Another test involved a hollow fiber membrane with raw water flow of 1.25 gpm, 170 to 200 psi, and 40 to 50% water recovery. The spiral-wound membrane system showed a 25% mercury removal, while the hollow fiber system removal efficiency was 79 to 81%.

Several preliminary ion-exchange experiments have been carried out by Ebersole and O'Connor (1972) to investigate organic and inorganic mercury removal from drinking water. These studies showed that as much as 98% of inorganic mercury added to distilled water could be removed by cation- and anion-exchange resins operated in series. Although these experiments were very preliminary, the results indicated that ion exchange may be an effective method for inorganic mercury removal.

References

Andres P (1984) IgA-IgG disease in the intestine of Brown Norway rats ingesting mercuric chloride. Clin Immunol Immunopathol 30 488–494.

Bernaudin JF, Druet E, Druet P, Masse R (1981) Inhalation or ingestion of organic or inorganic mercurials produces auto-immune disease in rats. Clin Immunol Immunopathol 20: 129–135.

Clarkson, TW (1972) The pharmacology of mercury compounds. Ann Rev Pharmacol 12: 375–406.

Druet P, Druet E, Potdevin F, Sapin C (1978) Immune type of glomerulonephritis induced by $HgCl_2$ in the brown Norway rat. Ann Immunol 129C: 777–792.

Ebersole G, O'Connor JT (1972) The removal of mercury from water by conventional water treatment processes. Presented at 92nd annual conference, American Waterworks Association, Chicago, IL, June.

Fitzhugh OG, Nelson A, Laug E, Junze F (1950) Chronic oral toxicants of mercuric-phenyl and mercuric salts. Arch Ind Occup Med 2: 433–442.

Gale TF (1974) Embryopathic effects of different routes of administration of mercuric acetate in the hamster. Environ Res 8:207–213.

Gleason MN, Gosselin RE, Hodge HC (1957) Clinical toxicology of commercial products. Williams and Wilkins Co, Baltimore, MD, p 154.

Jugo S (1976) Retention and distribution of $^{203}HgCl_2$ in suckling and adult rats. Hlth Physics 30:240–241.

Lehman AJ (1959) Appraisal of the safety of chemicals in foods, drugs, and cosmetics. Q Bull Assoc Food Drug Offic US.

Logsdon GS, Symons JM (1973) Mercury removal by conventional water treatment techniques. J Am Waterworks Assoc 65(8):554–562.

Pesce AJ, Hanenson I, Sethi K (1977) B_2 microglobulinuria in a patient with nephrotoxicity secondary to mercuric chloride ingestion. Clin Toxicol 11(3):309–315.

Rahola T, Hattula T, Korlainen A, Miettinen JK (1971) The biological halftime of inorganic mercury (Hg^{2+}) in man. Scand J Clin Invest 27(Suppl. 116):77.

Rothstein A, Hayes AD (1960) The metabolism of mercury in the rat studied by isotope techniques. J Pharmacol 130:166–176.

Seiler HG, Sigel H, Sigel A (eds) (1988) Handbook on the toxicity of inorganic compounds. Marcel Dekker, Inc, New York, NY.

Sigworth ES, Smith SB (1972) Adsorption of inorganic compounds by activated carbon. J Am Waterworks Assoc 64(6):386–391.

Sorg TJ (1977) Manual of treatment techniques for meeting the interim primary drinking water standards. USEPA Report no. EPA-600/8-77-005a. USEPA Office of Drinking Water, Washington, DC.

Sorg TJ (1979) Treatment technology to meet the interim primary drinking water regulations for organics: Part 4. J Am Waterworks Assoc 71:454–466.

Task Group on Metal Accumulation (1973) Accumulation of toxic metals with special reference to their absorption, excretion and biological halftimes. Environ Phys Biochem 3:65–67.

Theim L, Badorek D, O'Connor JT (1976) Removal of mercury from drinking water using activated carbon. J Am Waterworks Assoc. 68:447–451.

USEPA[a] (1973) Water quality criteria, 1972. Ecol Res Ser Rep of Water Quality Criteria. Report no. EPA-R3/73/033. USEPA Office of Water Regulation and Standards (OWRS), Washington, DC.

USEPA[a] (1979a) Method 245.1. Manual cold vapor technique. In: Methods for chemical analysis of water and wastes. Report no. EPA-600/4-79-020. USEPA Office of Drinking Water, Washington, DC.

USEPA[a] (1979b) Method 245.2. Automated cold vapor technique. In: Methods for chemical analysis of water and wastes. Report no. EPA-600/4-79-020. USEPA Office of Drinking Water, Washington, DC.

USEPA[a] (1980) Ambient water quality criteria for mercury. Report no. EPA-440/5-80-05b. USEPA Office of Water Regulations and Standards, Washington, DC.

USEPA[a] (1985) Drinking water criteria document for mercury (draft report). USEPA

[a] USEPA documents can be obtained by writing to the appropriate office within the U.S. Environmental Protection Agency, 401 M Street, S W., Washington, DC 20460. They may also be available by contacting the National Technical Information Service, 5285 Port Royal Road, Springfield, VA 22161 (Telephone 1-800-336-4700).

Office of Drinking Water (ODW), Washington, DC.
USEPA (1986) Guidelines for carcinogen risk assessment. Fed Regis 51(185):33992–34003. September 24.
USEPA[a] (1987) Estimated national occurrence and exposure to mercury in public drinking water supplies. Criteria and Standards Division, USEPA Office of Drinking Water (ODW), Washington, DC.
Weast RC (ed) (1971) CRC handbook of chemistry and physics, 52nd Ed. The Chemical Rubber Co, Cleveland, OH.
WHO (1971) World Health Organization. International standards for drinking water. WHO, Geneva, Switzerland.

Nickel[a]

I. General Information

Nickel occurs in both ionic and metallic forms. Nickel chloride is also known as nickelous chloride; nickel oxide is also called bunsenite.

Although this document is concerned with the toxic effects of ionic nickel, metallic nickel has the most uses, such as in the manufacture of stainless steel and various other alloys and in electroplating. The properties of nickel compounds are unique to the specific compound. Properties of nickel, nickel chloride, and nickel oxide are given in Tables 1, 2, and 3, respectively.

Nickel is a metallic element that is not found free in nature but exists as a number of salts. Nickel compounds are found in most geologic materials at levels up to several thousand ppm. Nickel occurs at low levels in most surface and ground waters. Nickel compounds are relatively insoluble, and therefore the levels of nickel in most surface and ground waters are generally less than 100 ppb. Because nickel compounds are used commercially in a number of industries, contamination of drinking water is the result of naturally occurring nickel compounds proliferated during industrial activities (USEPA 1979a; 1983a).

Nickel is a component of some plumbing materials. When pipes and other materials corrode, nickel can be released to drinking water. However, available information suggests that releases from this source are small (USEPA 1979a; 1983a).

Survey data on the occurrence of nickel in drinking water are limited. Based on these data, most supplies contain <40 μg/L of nickel. The highest level reported for a drinking water supply was 490 μg/L. Nickel also occurs at low levels in food. Based on the limited information available, diet is the major source of nickel exposure, with water making only a minor contribution (USEPA 1979a; 1983a).

II. Pharmacokinetics

The major routes of nickel intake for humans and animals are inhalation and ingestion and, to a lesser extent, percutaneous absorption. The extent of nickel absorption is dependent not only on the concentration of inhaled or ingested

[a] Based on USEPA (1985), except for Section I which is based on USEPA (1983a)

Table 1. General information and properties of nickel

Property	Value
Chemical Abstracts Service (CAS) number	7440-02-0
Chemical formula	Ni
Molecular weight	58.71
Physical state	Silver metal
Boiling point	2,732°C
Melting point	1,453°C
Density	8.90
Vapor pressure	—
Specific gravity	8.9
Water solubility	Insoluble
Log octanol/water partition coefficient	—
Odor threshold	—
Taste threshold	—
Air concentration conversion factor	—

Source Weast (1971)

Table 2. General information and properties of nickel chloride

Property	Value
Chemical Abstracts Service (CAS) number	7718-54-9
Chemical formula	$NiCl_2$
Molecular weight	129.62
Physical state	Yellow solid
Boiling point	973°C (sublimes)
Melting point	1,001°C
Density	3.55
Vapor pressure	—
Specific gravity	3.55
Water solubility	64.2 g/100 mL (20°C)
Log octanol/water partition coefficient	—
Odor threshold	—
Taste threshold	—
Air concentration conversion factor	—

Source Weast (1971)

Table 3. General information and properties of nickel oxide

Property	Value
Chemical Abstracts Service (CAS) number	1313-99-1
Chemical formula	NiO
Molecular weight	74.71
Physical state	Green-black solid
Boiling point	—
Melting point	1,990°C
Density	6.67
Vapor pressure	—
Specific gravity	6.67
Water solubility	—
Log octanol/water partition coefficient	—
Odor threshold	—
Taste threshold	—
Air concentration conversion factor	—

Source: Weast (1971)

nickel but also on its chemical and physical forms (USEPA 1985). Because inhalation and percutaneous exposures are not relevant to drinking water, emphasis here is placed on studies using oral exposure.

A. Absorption

Very little nickel ingested in food is absorbed. Total dietary intake ranges from 107 to 900 μg/day with average values of 160 to 500 μg daily (USEPA 1985); about 1 to 10% of this is absorbed (Horak and Sunderman 1973). In rats, intubation of ^{63}Ni in dilute acid solutions resulted in 3 to 6% absorption (Ho and Furst 1973). There was no uptake of nickel in rats chronically exposed to levels of 5 ppm in drinking water over the lifetime of the animal (Schroeder et al. 1974).

Absorption from inhalation exposure to nickel carbonyl is both rapid and extensive. Sunderman and Selin (1968) exposed rats to nickel carbonyl at 100 mg Ni/L of air for 15 min. It was estimated that half of the inhaled amount was initially absorbed. On the other hand, inhalation exposure to insoluble particulate nickel (e.g., the oxide or the subsulfide) results in little absorption.

B. Distribution

Tissue distribution in animals after oral exposure to Ni is concentration dependent. Calves fed supplemental nickel in the diet at levels of 62.5, 250, or 1,000 ppm showed somewhat elevated levels in pancreas, testes, and bone at

250 ppm; pronounced increases were seen in these tissues at 1,000 ppm (O'Dell et al. 1971).

Weanling rats fed dietary nickel acetate up to 1,000 ppm showed increased levels of nickel in kidney, liver, heart, and testes as concentration increased. The greatest accumulation was seen in the kidneys (Whanger 1973).

Transplacental transfer of nickel to the fetus takes place in both humans and animals. Newborn rats of mothers fed 1,000 ppm Ni in the diet showed whole-body levels of 22 to 30 ppm (Phatak and Patwardhan 1950). Also, levels of nickel (0.04 to 2.8 ppm) seen in the liver, heart, and muscle of fetuses were similar to those seen in adult humans (Casey and Robinson 1978).

C. Metabolism

Serum albumin is the main carrier protein for nickel in the sera of humans, rabbits, rats, and bovine species. In the sera of rabbits and humans, the nickel-rich metalloproteins α_1-macroglobulin (nickeloplasmin) and 9.5 S α_1-glycoprotein, respectively, have been identified (NAS 1975).

D. Excretion

The main excretory route of absorbed nickel in humans and animals appears to be urine (Ho and Furst 1973); biliary excretion is also observed in experimental animals (Onkelinx et al. 1973). Deposition in human hair also appears to be an excretory mechanism (Nechay and Sunderman 1973). Unabsorbed dietary nickel is excreted in feces.

III. Health Effects

A. Humans

No clinical or epidemiologic studies on the toxicity of nickel after oral exposure were found in the available literature. Its toxicity to humans and animals is a function of the chemical form and the route of exposure. A correlation between chronic inhalation exposure to nickel carbonyl and respiratory tract cancer was suggested by epidemiologic studies and confirmed in experimental animals. Dermatitis (nickel itch) is another frequent effect of exposure (USEPA 1983b).

B. Animals

1. Short-Term Exposure. The oral LD_{50} values converted to mg nickel/kg range from 105 mg/kg for nickel chloride in male rats to 186 mg/kg for nickelocene in mice (USEPA 1985).

Nickel chloride administered orally to rats at doses of 0.5 to 5.0 mg/kg/d for 2 to 4 wk led to a significantly decreased thyroid absorption of iodine (Lestrovoi et al. 1974).

Nickel acetate in the diet of weanling OSU brown rats for 6 wk at concentrations of 100, 500, or 1,000 ppm (i.e., 10, 50, or 100 mg Ni/kg) resulted in a significantly reduced weight gain at 500 ppm; rats exposed to 1,000 ppm lost weight. At 500 and 1,000 ppm, there was a dose-related decrease in blood hemoglobin concentration, packed-cell volume, and plasma alkaline phosphatase activity. Cytochrome oxidase activity was decreased significantly ($p < 0.005$) in both heart and liver in the high-dose group. Iron concentration was increased significantly ($p < 0.05$) in red blood cells, heart, kidney, liver, and testes in the 1,000-ppm group; elevated levels of iron concentration also were seen in the 500-ppm group. No significant effects were seen on body weight, mineral content, and enzyme activity in the 100-ppm group in comparison with levels in control. The 100 ppm (10 mg Ni/kg) level is considered a No-Observed-Adverse-Effect Level (NOAEL); 500 ppm (50 mg Ni/kg) is a Lowest-Observed-Adverse-Effect Level (LOAEL) (Whanger 1973).

2. Long-Term Exposure. Nickel added to the diet of mice resulted in reduced body weight gain in females at a dietary concentration of 1,100 ppm nickel and reduced body weight gain in both males and females at 1,600 ppm (Weber and Reid 1969).

Nickel (as nickel chloride) administered to rats at 225 ppm in drinking water (17.6 mg Ni/kg) for 4 mon led to a significant reduction in body weight ($p < 0.05$) compared with controls (Clary 1975). Daily urinary volume and urinary zinc and calcium concentrations were reduced significantly. Also, at sacrifice, serum lipid and cholesterol concentrations were significantly reduced ($p < 0.05$).

Daily doses of 25 mg/kg of nickel sulfate administered by oral intubation to male rats for 120 d caused degenerative cellular changes in the liver and kidney (von Waltschewa et al. 1972). In treated rats, testes were smaller than in controls. Other testicular changes included interstitial cell proliferation, transparent vessel walls, reduced numbers of spermatozoa and their precursors, and decreased concentrations of succinodehydrogenase and steroid 3-β-dehydrogenase.

Rats fed a diet containing nickel acetate at concentrations of 0.1 to 1.0% (16 to 160 mg Ni/kg) for 10 to 190 d showed a high rate of mortality; hypoplasia of bone marrow, thymus, and spleen; progressive renal tubular degeneration; mural exudative pulmonary alveolar lesions; and noninflammatory lysis of pancreatic exocrine cells (Ashrof and Sybers 1974).

In a chronic study with mice fed a diet devoid of cadmium and low in other metals with 5 ppm nickel added to their drinking water (approximately 0.85 mg Ni/kg/d) no significant effects were observed. Only body weights of animals dying after 1 yr were depressed 4% to 13% (Schroeder et al. 1964).

Mean body weights of both male and female rats were reduced significantly ($p < 0.025$) at 18 mon in a study in which rats were administered 5 ppm nickel (average daily dose estimated to be 0.41 mg Ni/kg) in drinking water for life (Schroeder et al. 1974). Lifespan was not affected. Histopathologic studies revealed an increased incidence (13.3%; $p < 0.025$) of focal myocardial fibrosis in the experimental group.

In a 2-yr feeding study with beagle dogs fed nickel sulfate hexahydrate at 0, 100, 1,000, or 2,500 ppm (0, 3, 29, or 70 mg Ni/kg), no significant effects on body weight, hematologic values, results of urinalysis, organ-to-body weight ratios, or histopathologic findings were noted at 100 or 1,000 ppm. At 2,500 ppm, body weight gain was depressed, hemoglobin and hematocrit values tended to be lower, and kidney-to-body weight and liver-to-body weight ratios were significantly higher ($p < 0.05$). Pathologic changes in the lungs and granulocytic hyperplasia of bone were also noted in the high-dose group. The NOAEL from this study is 1,000 ppm (29 mg/kg) (Ambrose et al. 1976).

In a 2-yr feeding study in rats given 0, 100, 1,000, or 2,500 ppm nickel sulfate in milk (0, 5, 50, and 125 mg Ni/kg), no significant effects were reported at 100 ppm (Ambrose et al. 1976). Body weight was significantly reduced ($p < 0.05$) in both male and female rats fed 2,500 ppm. At 1,000 ppm, body weight was also reduced in both sexes. Animals fed 1,000- or 2,500-ppm diets had significantly higher ($p < 0.05$) heart-to-body weight ratios and significantly lower liver-to-body weight ratios. The 1,000-ppm (50 mg/kg) dose represents a LOAEL from this study, and 100 ppm (5 mg/kg) is a NOAEL.

3. Reproductive Effects. In a three-generation reproduction study in rats, nickel sulfate hexahydrate fed at 0, 250, 500, or 1,000 ppm (0, 12.5, 25, or 50 mg Ni/kg/d) led to a slight decrease in adult body weight at mating and weaning in the 1,000-ppm group. Fertility, gestation, viability, and lactation indices were not affected. The body weights of weanlings from the 1,000 ppm group were reduced in all generations. The incidence of stillborn pups was 19%, 12%, and 15% in the F_{1a} and 4%, 20%, and 25% in the F_{1b} in the 250-, 500-, and 1,000-ppm groups, respectively, compared with 4% and 2% in the control F_{1a} and F_{1b}. Increased incidence of fetal mortality was not observed in the F_2 and F_3 (Ambrose et al. 1976).

In another three-generation reproduction study, rats were provided drinking water containing 5 ppm nickel (salt not specified, estimated total daily dose was 0.43 mg/kg) (Schroeder and Mitchener 1971). Neonatal mortality was increased significantly ($p < 0.025$) in all generations of exposed rats compared with controls. The number of runts was increased significantly in the F_1 ($p < 0.025$) and F_3 ($p < 0.0001$) generations. Average litter size was reduced somewhat in the F_3. The diet was found to be deficient in trace metals, particularly chromium.

No significant differences were observed in the litter size and initial body weight of pups when male and female rats were fed diets containing 250, 500, or 1,000 ppm nickel (daily dose of 10, 20, or 40 mg Ni/kg) for 8 wk before breeding and continuing through lactation (Phatak and Patwardhan 1950).

4. Developmental Effects. Transplacental distribution of nickel is well documented in laboratory animals (USEPA 1985).

In a three-generation reproduction study in rats, which was described previously, no evidence of teratogenicity was seen in weanlings of rats fed nickel sulfate hexahydrate at levels of 0, 250, 500, or 1,000 ppm (0, 12.5, 25, or 50 mg Ni/kg/d) (Ambrose et al. 1976).

5. Mutagenicity. Nickel chloride was not mutagenic in *Escherichia coli* and *Bacillus subtilis* (USEPA 1985). Nickel chloride and nickel sulfate were mutagenic or weakly mutagenic in eukaryotic test systems (USEPA 1985). Nickel-induced chromosomal aberrations have been observed in cultured mammalian cells. Nickel-induced sister-chromatid exchanges have been observed in both cultured mammalian cells and in human lymphocytes (USEPA 1985).

6. Carcinogenicity. Statistically significant elevations in the incidence of respiratory tract cancers have been demonstrated in nickel refinery workers (IARC 1976; NAS 1975; NIOSH 1977); these data are not, however, relevant to the consumption of nickel in drinking water.

Repeated intraperitoneal injections of nickel acetate at a dose of 360 mg/kg have induced lung carcinomas in mice (Stoner et al. 1976). This is not, however, relevant to the consumption of nickel in drinking water. No evidence of carcinogenicity has been found in those chronic studies in which nickel was administered orally to laboratory animals (Schroeder et al. 1964, 1974; Schroeder and Mitchener 1975).

IV. Quantification of Toxicological Effects[a]

A. One-Day Health Advisory

The available data are insufficient to develop a One-Day Health Advisory (HA) for nickel. It is recommended that the Ten-Day HA of 1.0 mg/L be used as the One-Day HA for a 10-kg child.

B. Ten-Day Health Advisory

The study by Whanger (1973) was selected for the derivation of a Ten-Day HA. Dose–response relationships were observed in this 6-wk dietary study

[a] See Introduction for discussion of methodology

identifying a NOAEL for nickel of 100 ppm in diet (10 mg/kg/d) and a LOAEL of 500 ppm in diet (50 mg/kg/d). The biological end points included body weight gain, hematologic parameters, and cytochrome oxidase activity. The Ten-Day HA for Ni for a 10-kg child is calculated as follows:

$$\text{Ten-Day HA} = \frac{(10 \text{ mg/kg/d})(10 \text{ kg})}{(100)(1 \text{ L/d})} = 1.0 \text{ mg/L } (1{,}000 \ \mu\text{g/L})$$

C. Longer-Term Health Advisory

The available data are insufficient to develop a Longer-Term HA for nickel. It is recommended that the Drinking Water Equivalent Level (DWEL) of 0.58 mg/L be used as the Longer-Term HA for a 70-kg adult and the modified DWEL of 0.16 mg/L (adjusted for a 10-kg child) be used as the Longer-Term HA for a 10-kg child.

The USEPA is in the process of reviewing a draft report of a 90-d gavage study in rats (Mayhew 1987). After the official final report has been reviewed and considered, it may serve as the basis for a Longer-Term HA.

D. Lifetime Health Advisory

Because of various problems with the teratogenicity and reproductive toxicity studies of Schroeder and Mitchener (1971) and Ambrose et al. (1976), the 2-yr rat feeding study of Ambrose et al. (1976) was selected for use in determining the Lifetime HA. Rats were given 0, 100, 1,000, or 2,500 ppm dietary nickel sulfate (approximate daily dose was 0, 5, 50, or 125 mg Ni/kg). The study identified a NOAEL of 100 ppm (5 mg/kg/d). Using this NOAEL, the Lifetime HA is derived as follows:

Step 1: Determination of the Reference Dose (RfD)

$$\text{RfD} = \frac{(5 \text{ mg/kg/d})}{(100)(3)} = 0.016 \text{ mg/kg/d}$$

where 100 = uncertainty factor, chosen in accordance with NAS/ODW guidelines for use with a NOAEL from an animal study; and 3 = additional uncertainty factor used because of inadequacies in the reproductive studies.

Step 2: Determination of the Drinking Water Equivalent Level (DWEL)

$$\text{DWEL} = \frac{(0.016 \text{ mg/kg/d})(70 \text{ kg})}{(2 \text{ L/d})} = 0.58 \text{ mg/L } (580 \ \mu\text{g/L})$$

Step 3: Determination of the Lifetime HA

The Lifetime HA is calculated as follows:

$$\text{Lifetime HA} = 0.58 \times 30\% = 0.17 \text{ mg/L } (200 \ \mu\text{g/L})$$

E. Evaluation of Carcinogenic Potential

Oral exposure to nickel has not been shown to be carcinogenic. Data are not available concerning the potential carcinogenic effects of ingested nickel compounds in humans.

There is evidence, however, that certain nickel compounds or mixtures of nickel compounds are carcinogenic in humans via inhalation. Nickel refinery dust and nickel subsulfide, which is believed to be the major nickel component of refinery dust, are classified in Group A: human carcinogen, based on USEPA guidelines for assessment of carcinogen risk (USEPA 1986). Nickel carbonyl has been classified as Group B2: probable human carcinogen, based on sufficient evidence of carcinogenicity in animals, but insufficient epidemiologic data in humans.

Based on an evaluation of the carcinogenic potential of nickel from inhalation and intramuscular injection, the International Agency for Research on Cancer (IARC) classified nickel and certain nickel compounds in Group 2A (IARC 1976). Currently, however, evidence is insufficient to classify nickel as a carcinogen after oral exposure.

V. Other Criteria, Guidance, and Standards

The American Conference of Governmental Industrial Hygienists (1983) has established a Time-Weighted Average (TWA) and Threshold Limit Value (TLV) of 1.0 mg Ni/m^3 for metallic nickel salts and 0.1 mg Ni/m^3 for soluble nickel salts.

The National Institute of Occupational Safety and Health (1977) criterion for occupational exposure to nickel is a TWA of 15 μg Ni/m^3.

The USEPA (1980; 1982) derived an Acceptable Daily Intake of 1.46 mg Ni/d and established an ambient water quality criterion of 0.632 mg Ni/L.

VI. Analytical Methods

Nickel levels are determined by atomic absorption (AA) using either direct aspiration into a flame (USEPA 1979b) or a furnace technique (USEPA 1979c).

The direct-aspiration AA procedure is a physical method based on nickel's absorption of radiation at 232.0 nm. The sample is aspirated into an air-acetylene flame and atomized. A light beam is directed through the flame into a monochromator and onto a detector that measures the amount of light absorbed. Absorbance is proportional to the concentration of nickel in the sample. The detection limit is 40 μg/L.

The furnace AA procedure is similar to direct-aspiration AA, except a furnace rather than a flame is used to atomize the sample. Detection limit is 1 μg/L.

VII. Treatment Technologies

Treatment techniques capable of removing nickel from drinking water include lime softening, ion exchange, and reverse osmosis. Conventional coagulation is moderately effective in removing nickel from drinking water. Although its removal from drinking-water supplies by these technologies has not been studied extensively, some information is available from wastewater technology surveys.

Culp et al. (1978) reported excellent removal (90.9 to 99.9%) of nickel with lime softening for wastewater with concentrations of 5 mg/L to 160 mg/L. Maruyama et al. (1975) reported removal efficiencies of 95% with low lime softening (260 mg/L lime dosage) and 98% with high lime softening (600 mg/L lime dosage) from domestic wastewater containing 5 mg/L.

Cation exchange has been used extensively in the plating industry to recover nickel. Usually, these operations have employed cation resins in the hydrogen cycle because of the need to recover both acid and metal for recycling. Nickel was eluted with sulfuric acid, 6 to 10 lb/ft^2 resin (29–49 kg/m^2 resin) in 10% solution. The reported efficiencies for removal from plating industry wastewater are 96 to 100% (Keramida and Etzel 1982).

Reverse-osmosis membranes remove nickel effectively from source water. A laboratory-scale study evaluating the performance of cellulose acetate membrane with plating rinse showed that cellulose acetate has a rejection efficiency for Ni^{2+} of 99.6%. Other commercially available membranes are cellulose acetate butyrate, nylon hollow fibers, and polyurethanes (Golomb 1972). These membranes, however, have not been tested for their efficiencies in nickel removal. The cellulose acetate membrane was field tested on a small industrial automatic plating line. The wastewater nickel concentration varied (1,700 mg/L, 50 mg/L, and 12 mg/L). Tests by Golomb (1974) showed that cellulose acetate membrane can be used to remove effectively 99 + % of nickel from waste-rinse streams.

Pilot-plant studies evaluating the efficiency of coagulation indicated that alum was only 25 to 45% effective in the removal of nickel from water (Maruyama et al. 1975; Hannah et al. 1977).

Another study by Nillson (1978) investigated the removal of trace metals from tap water and municipal wastewater. Calcium hydroxide was 91.3% effective in removing nickel from tap water and 63.3% effective in removal from wastewater at a pH of 9.5.

References

ACGIH (1983) American Conference of Governmental Industrial Hygienists. TLVs—threshold limit values for chemical substances and physical agents in the work environment with intended changes for 1983–84. ACGIH, Cincinnati, OH, p 27

Ambrose AM, Larson PS, Borzelleca JR, Hennigar GR Jr (1976) Long-term toxicologic assessment of nickel in rats and dogs. J Food Sci Technol 13: 181–187.

Ashrof M, Sybers HD (1974) Lysis of pancreatic exocrine cells and other lesions in rats fed nickel acetate. Am J Pathol 74: 102a.

Casey CE, Robinson MF (1978) Copper, manganese, zinc, nickel, cadmium, and lead in human foetal tissues. Brit J Nutr 39: 639–646.

Clary JJ (1975) Nickel chloride-induced metabolic changes in the rat and guinea pig. Toxicol Appl Pharmacol 31: 55–65.

Culp RJ, Wesner GM, et al. (1978) Handbook of advanced wastewater treatment, 2nd Ed. Van Nostrand Reinhold Co., New York, NY.

Golomb A (1972) Application of reverse osmosis to electroplating waste treatment. Plating 59(4): 316–319.

Golomb A (1974) Application of reverse osmosis to electroplating waste treatment. Plating 61(5): 432–442.

Hannah SA, Telus M, Cohen JM (1977) Removal of uncommon trace metals by physical and chemical treatment processes. J Water Pollut Control Fed 49(11): 2297–2309.

Ho W, Furst A (1973) Nickel excretion by rats following a single treatment. Proc West Pharmacol Soc 16: 245–248.

Horak E, Sunderman FW Jr (1973) Fecal nickel excretion by healthy adults. Clin Chem 19: 429–430.

IARC (1976) International Agency for Research on Cancer. Nickel and nickel compounds. IARC monographs 2: 75–112. IARC, Lyon, France.

Keramida V, Etzel JE (1982) Treatment of metal plating wastewater with a disposable ion exchange material. In: Proceedings of the 37th industrial waste conference. Purdue University, Lafayette, IN.

Lestrovoi AP, Itskova AI, Eliseev IN (1974) Effect of nickel on the iodine fixation of the thyroid gland when administered perorally and by inhalation. Gig Sanit 10: 105–106.

Maruyama T, Hannah SA, Cohen JM (1975) Metal removal by physical and chemical treatment processes. J Water Pollut Control Fed 47(5): 962–975.

Mayhew DA[a] (1987) Ninety-day gavage study in albino rats using nickel. Draft final report by American Biogenics Corp, Decatur, IL. USEPA Office of Drinking Water, Washington, DC.

NAS (1975) National Academy of Sciences. Nickel. National Academy of Sciences committee on medical and biological effects of environmental pollutants. NAS, Washington, DC.

Nechay MW, Sunderman FW Jr (1973) Measurements of nickel in hair by atomic absorption spectrometry. Ann Clin Lab Sci 3: 30–35.

Nillson R (1978) Removal of metals by chemical treatment of municipal wastewater. Water Res 5: 51–60.

NIOSH (1977) National Institute of Occupational Safety and Health. Criteria for a recommended standard ... occupational exposure to inorganic nickel. NIOSH Publication no. 77-164. NIOSH, Washington, DC.

O'Dell GD, Miller WJ, King A, Moore SL, Blackmon DM (1971) Effect of dietary nickel level on excretion and nickel content of tissues in male calves. J An Sci 32: 769–773.

Onkelinx C (1973) Compartmental analysis of the metabolism of $^{63}Ni(II)$ in rats and rabbits. Res Comm Chem Pathol Pharmacol 6:663.

Phatak SS, Patwardhan VN (1950) Toxicity of nickel. J Sci Indian Res 9B:70–76.

Schroeder HA, Balassa JJ, Vintin WH Jr (1964) Chromium, lead, cadmium, nickel, and titanium in mice: effect on mortality, tumors and tissue levels. J Nutr 83:239–250.

Schroeder HA, Mitchener M (1971) Toxic effects of trace elements on the reproduction of mice and rats. Arch Environ Hlth 23:102–106.

Schroeder HA, Mitchener M, Nason AP (1974) Life-term effects of nickel in rats: survival, tumors, interactions with trace elements and tissue levels. J Nutr 104:239–243.

Schroeder HA, Mitchener M (1975) Life-term effects of mercury, methyl mercury and nine other trace metals on mice. J Nutr 105:452–458.

Stoner GD, Shimkin MB, Troxell MC, Thompson TL, Terry LS (1976) Test for carcinogenicity of metallic compounds by the pulmonary tumor response in strain A mice. Cancer Res 36:1744–1747.

Sunderman FW Jr, Selin CE (1968) The metabolism of nickel-63 carbonyl. Toxicol Appl Pharmacol 12:207.

USEPA[a] (1979a) Water related environmental fate of 129 priority pollutants. EPA Report no. EPA-440/4-79-029. USEPA Office of Water Planning and Standards, Washington, DC.

USEPA[a] (1979b) Method 249.1. Atomic absorption, direct aspiration. In: Methods for chemical analysis of water and wastes. Report no. EPA-600/ 4-79-020. USEPA Environmental Monitoring and Research Laboratories, Cincinnati, OH.

USEPA[a] (1979c) Method 249.2. Atomic absorption, furnace technique. In: Methods for chemical analysis of water and wastes. Report no. EPA-600/4-79-020. USEPA Environmental Monitoring and Research Laboratories, Cincinnati, OH.

USEPA[a] (1980) Ambient water quality criteria document for nickel. EPA-440/4-80-060. NTIS[b] PB 81-117715. USEPA Environmental Criteria and Assessment Office (ECAO), Cincinnati, OH.

USEPA[a] (1982) Errata for ambient water quality criteria documents. February 23. p 14. USEPA Environmental Criteria and Assessment Office, Cincinnati, OH.

USEPA[a] (1983a) Nickel occurrence in drinking water, food, and air. USEPA Office of Drinking Water (ODW), Washington, DC.

USEPA[a] (1983b) Health assessment document for nickel. Report no. EPA-600/ 8-83-012. USEPA Office of Research and Development (ORD), Environmental Criteria and Assessment Office (ECAO). Research Triangle Park, NC.

USEPA (1985) Drinking water criteria document for nickel. Report no. EPA-600/

[a] USEPA documents can be obtained by writing to the appropriate office within the U.S. Environmental Protection Agency, 401 M Street, S W., Washington, DC 20460. They may also be available by contacting the National Technical Information Service, 5285 Port Royal Road, Springfield, VA 22161 (Telephone: 1-800-336-4700).

[b] Documents with an NTIS publication number are available from the National Technical Information Service, 5285 Port Royal Road, Springfield, VA 22161 (Telephone 1-800-336-4700).

X-84-193-1. Environmental Criteria and Assessment Office (ECAO), Cincinnati, OH.

USEPA (1986) Guidelines for carcinogen risk assessment. Fed Regis 51(185):33992–34003. September 24.

Von Waltschewa W, Slatewa M, Michailow I (1972) Hodenveranderungen bei weissen Ratten durch chronische Verabreichung von Nickel sulfat. (Testicular changes due to long-term administration of nickel sulfate in rats.) Exp Pathol 6:116–120. (Ger. with Eng. Abstract)

Weast RC (ed) (1971) CRC handbook of chemistry and physics, 52nd Ed. The Chemical Rubber Co, Cleveland, OH.

Weber CW, Reid BL (1969) Nickel toxicity in young growing mice. J An Sci 28:620–623.

Whanger PD (1973) Effects of dietary nickel on enzyme activities and mineral content in rats, Toxicol Appl Pharmacol 25:323–331.

Nitrate and Nitrite[a]

I. General Information

Nitrate and nitrite have a variety of commercial uses. The major use of nitrate is in inorganic fertilizers. It is also used in the manufacture of explosives, in glassmaking, and as a heat-transfer fluid and a heat-storage medium for solar heating applications. Both nitrate and nitrite are used in curing meats. The properties of potassium nitrate and nitrite, two commonly used compounds, are shown in Table 1 (USEPA 1985). Potassium nitrate is also known as saltpeter (Windholz 1976).

Nitrate and nitrite are naturally occurring inorganic ions that make up part of the nitrogen cycle. Wastes containing organic nitrogen enter the soil and are decomposed to ammonia, which is subsequently oxidized to nitrite and nitrate. Because nitrite is easily oxidized to nitrate, nitrate predominates in ground and surface waters. Nitrate is then taken up by plants during their growth and converted back to organic form. Levels of nitrate in water can be increased through contamination by nitrogen-containing fertilizers or human and animal wastes. Nitrate and nitrite ions are mobile in soil and move steadily with ground water (USEPA 1987).

Surveys of naturally occurring levels of nitrate and nitrite in ground and surface water revealed that levels normally do not exceed 1 to 2 mg/L for nitrate and 0.1 mg/L for nitrite. Surface waters usually contain lower levels of nitrate and nitrite than ground water. Nitrate has been included in a number of drinking water surveys, which indicate its occurrence at levels of less than 1 mg/L in most surface and ground water supplies. Nitrates occur at levels exceeding 5 mg/L in about 3% of surface waters and 6% of ground waters. Currently, 40 surface water supplies and 568 ground-water supplies are known to exceed the nitrate Maximum Contaminant Level (MCL) of 10 mg/L. Systems that exceed the MCL are usually contaminated by nitrates as a result of the use of fertilizers or from animal wastes or septic systems. Nitrite levels have not been surveyed in drinking water supplies but are expected to be much lower than 1 mg/L (USEPA 1987).

Nitrates occur naturally in a number of foods, particularly vegetables. They are also added to meat products as preservatives. For adults, the major

[a] Based on USEPA (1985), except for Section I which is based on USEPA (1987).

Table 1. General information and properties of potassium nitrate and potassium nitrite

Property	Value: Potassium nitrate	Value: Potassium nitrite
Chemical Abstracts Service (CAS) number	7757-79-1	7758-09-0
Chemical formula	KNO_3	KNO_2
Molecular weight	101.11	85.11
Physical state	Solid	Solid
Boiling point	NA (decomposes)	NA (decomposes)
Melting point	334°C	440°C
Density (16°C)	2.109	1.915
Vapor pressure	—	—
Specific gravity	2.1	1.9
Water solubility	13.3 g/100 mL (0°C)	281 g/100 mL (0°C)
Log octanol/water partition coefficient	—	—
Odor threshold	—	—
Taste threshold	—	—

Source: Weast (1974)

source of nitrates is dietary. For infants, water appears to be the major source (USEPA 1987).

II. Pharmacokinetics

A. Absorption

Nitrate is absorbed by active transport from the upper small intestine, and nitrite is absorbed by diffusion across both the gastric mucosa and the wall of the intestine (USEPA 1985).

Both nitrate and nitrite are readily and completely absorbed after oral administration. Parks et al. (1981) reported that both $^{13}NO_3^-$ and $^{13}NO_2^-$ were completely absorbed within 10 min after administration of 10 to 100 mg/kg in mice (Parks et al. 1981). Similar results for nitrate (dose not specified) in rats were reported by Witter et al. (1979).

B. Distribution

Both nitrate and nitrite readily distribute throughout the tissues but do not bioaccumulate. Rapid, homogeneous distribution of nitrate (dose unspecified) was observed in rats 45 to 60 min after dosing by gavage (Witter et al. 1979).

Both $^{13}NO_3^-$ and $^{13}NO_2^-$ achieved transient equilibrium in mice within 5 min after intratracheal administration of 10 to 100 mg/kg (Parks et al. 1981). Equilibrium between the intravascular and extravascular compartments of rabbits was reached within 5 min after injection of either radiochemical.

Nitrate secretion in saliva by humans was reported by Spiegelhalder et al. (1976) after ingestion of vegetables and vegetable juices. Secretion of nitrate by the gastric mucosa in rats was observed by Bloomfield et al. (1962) after intraperitoneal doses of sodium nitrate ranging from 60 to 200 mg/kg.

In rats, nitrite crosses the placenta (Shuval and Gruener 1977).

No evidence has been found for bioaccumulation of nitrate or nitrite in any tissue (USEPA 1985).

C. Metabolism

Nitrate is not directly metabolized to other compounds in humans. However, bacterial metabolism of nitrate in humans, particularly infants, can markedly decrease the ability of blood to carry oxygen to tissues.

1. Humans. Although no evidence exists that mammals metabolize nitrate into other compounds (Parks et al. 1981), bacteria found in human saliva and the stomach can reduce nitrate to nitrite (USEPA 1985).

Due to decreased acidity (increased pH), particularly in the stomach of bottle-fed infants, bacteria capable of reducing nitrate to nitrite may proliferate in the stomach, thus leading to an increased formation of nitrite in infants 3 mon old or younger (USEPA 1985).

Nitrite reacts with hemoglobin (the protein responsible for the ability of blood to transport oxygen to tissues) in erythrocytes to form methemoglobin, which is incapable of transporting oxygen (Parks et al. 1981).

The enzyme methemoglobin reductase converts methemoglobin to hemoglobin and nitrate, thus reversing the process induced by nitrite (Smith and Beutler 1966).

Bacteria in saliva reduced 5% of absorbed nitrate into nitrite (Spiegelhalder et al. 1976).

2. Animals. Gruener et al. (1973) observed that the activity of methemoglobin reductase in rat fetuses was nearly 10 times higher than that in adult rats.

Other animal studies report that nitrite in the stomach can react with secondary amines and other amine substrates to form *N*-nitroso compounds that may be oncogenic (Ohshima and Bartsch 1981; Sander and Schweinsberg 1972). Vitamin C and vitamin E can inhibit formation of nitrosamines (Archer et al. 1975; Kamm et al. 1977).

D. Excretion

Nitrate is readily excreted by the kidneys (USEPA 1985). In humans, about 25% of absorbed nitrate is secreted in saliva (Spiegelhalder et al. 1976). Although it has been suggested that appreciable amounts of nitrate are eliminated in human USEPA 1985 and bovine milk (Davidson et al. 1964), data are inadequate to support this conclusion.

The half-life for elimination of nitrite in dogs, sheep, and Shetland ponies (0.5 to 0.6 hr) is too rapid to be attributed solely to renal excretion and suggests that metabolism may be significant (Schneider and Yeary 1975).

III. Health Effects

A. Humans

The lethal dose (LD) of potassium nitrate for an adult ranges from 54 to 462 mg/kg; the LD of sodium nitrate ranges from 32 to 154 mg/kg (Burden 1961).

The toxicity of nitrate in humans is due to reduction of nitrate to nitrite. By reacting with hemoglobin, nitrite forms methemoglobin, does not transport oxygen to the tissues, and thus leads to possible asphyxia (see Metabolism) (USEPA 1985).

Normal methemoglobin levels in humans range from 1 to 2% (Shuval and Gruener, 1977). A level greater than 3% is defined as methemoglobinemia. However, there is a consistent elevation of the methemoglobin concentration in pregnant women from the 14th gestational week through delivery (Skrivan 1971).

A survey by the American Public Health Association revealed that more than 278 cases of cyanosis in infants were associated with nitrate-contaminated water (Walton 1951). No cases of cyanosis in infants were associated with water containing 10 mg/L or less of nitrate-nitrogen (see Ten-Day Health Advisory below).

Winton et al. (1971) compared methemoglobin levels with nitrate ion intake in 111 infants younger than 6 mon. Only three infants had methemoglobin levels of more than 2.9%. They were the youngest of five infants who had received more than 10 mg/kg/d of nitrate ion.

Craun et al. (1981) conducted an epidemiologic study of 102 children aged 1 to 8 yr in Washington County, Illinois. Of the study subjects, 64 consumed water with high nitrate levels (22 to 111 mg/L nitrate-nitrogen), and 38 consumed water with low nitrate levels (<10 mg/L). Ingestion of water containing 22 to 110 mg/L did not produce abnormal mean methemoglobin levels and was not related to increased methemoglobin levels (see also Ten-Day HA).

Hegesh and Shiloah (1982) demonstrated that nitrites were synthesized in infants with acute diarrhea (see also Ten-Day Health Advisory).

In pregnant women, the level of methemoglobin increases from normal (0.5 to 2.5% of total hemoglobin) to a maximum of 10.5% at the 30th week of pregnancy followed by a subsequent decline to normal after delivery (Skrivan 1971). Thus, pregnant women may be more sensitive to the induction of clinical methemoglobinemia by nitrite at approximately the 30th wk of pregnancy.

B. Animals

1. Short-Term Exposure. In rabbits and rats, acute oral LD_{50} values for potassium nitrate of 1,166 mg/kg and 1,986 mg/km, respectively, were reported (Windholz 1976; WHO 1962). The acute oral LD_{50} of sodium nitrate in rabbits was 1,955 mg/kg (Windholz 1976).

In rats, reported acute oral LD_{50} values for sodium nitrate ranged from 46 to 120 mg/kg (Druckery et al. 1963; Imaizumi et al. 1980; Windholz 1976; WHO 1962).

Unlike the effect in humans, in whom nitrite toxicity is attributed to the formation of methemoglobin (see Metabolism), the immediate toxic effect of nitrite in some species (e.g., horses) is due to nitrite-induced vasodilation, which results in cardiovascular collapse and shock (USEPA 1985).

In a 3-wk drinking water study in mice, elevated methemoglobin levels were observed in 50-d-old mice given nitrite ion (as sodium nitrite) at 133 and 178 mg/kg/d; methemoglobin levels were not elevated at 88 mg/kg/d (Shuval and Gruener 1977).

2. Long-Term Exposure. In a 6-mon rat feeding study, 2,500 mg nitrate/kg/d produced a marked diuretic effect within 2 mon when compared with rats fed equimolar levels of sodium chloride; 250 mg nitrate/kg/d caused no diuresis and was identified as the No-Observed-Adverse-Effect Level (NOAEL) in this study (Fritsch et al. 1980).

In a 6-mon feeding study in rats, 250 and 2,500 mg nitrate/kg/d and 25 and 250 mg nitrite/kg/d induced hemorrhagic areas in the spleen (Fritsch et al. 1980). Therefore, 250 mg nitrate/kg/d and 25 mg nitrite/kg/d were identified as Lowest-Observed-Adverse-Effect Levels (LOAELs) in this study.

Two long-term studies using ICR mice reported increases in amyloidosis (starchy deposits) and hemosiderosis after ingestion of high doses of sodium nitrate (2,500 and 5,000 mg nitrate/kg/d) (Sugiyami et al. 1979) and sodium nitrite (208, 416, and 833 nitrite/kg/d) (Inai et al. 1979). LOAELs of 2,500 mg nitrate/kg/d and 208 mg nitrite/kg/d can be identified from the results of these studies.

3. Reproductive Effects. In a developmental toxicity study by Globus and Samuel (1978), no sodium nitrite-induced adverse reproductive effects were evidenced (see Developmental Effects).

4. Developmental Effects. Groups of mice were intubated and given sodium nitrite at 16.7 mg/kg/d on d 0 through 14, 16, or 18 of gestation (Globus and Samuel 1978). Analysis of fetal livers indicated that sodium nitrite administered to mothers stimulated fetal hepatic erythropoiesis. Nitrite-related effects on fetal mortality, resorptions, mean weight, number of offspring, or incidence of skeletal malformation were not observed.

The nitrosation of amides or amines in the stomach produces *N*-nitroso compounds that may pass through the placenta to exert teratogenic or fetotoxic effects (Ivankovic 1979; Teramoto et al. 1980).

5. Mutagenicity. Both sodium nitrite and sodium nitrate were negative in host-mediated assays in mice (FDA 1972a,b). Other host-mediated assays did not reveal sodium nitrite to be mutagenic in mice (Couch and Friedman 1975) or in either rats or mice (Whong et al. 1979).

Results of dominant lethal gene tests in rats were negative for both sodium nitrate and nitrite (USFDA 1972a,b); results of a cytogenetic assay in rat marrow cells were also negative for both compounds.

Kodama et al. (1976) reported that sodium nitrite induced mutations to azaguanine resistance in cultured FM3A cells (a C3H mouse mammary carcinoma cell line). Sodium nitrite was mutagenic in *Salmonella typhimurium* (USFDA 1972a,b) and *Escherichia coli* Sd-4 (Hussain and Ehrenberg 1974).

6. Carcinogenicity. The USEPA (1985) judged that the available animal studies (Newberne 1978, 1979; Maekawa et al. 1982) provided inconclusive evidence regarding the carcinogenicity of nitrate and nitrite administered orally in the absence of nitrosatable compounds.

Many studies have documented the carcinogenicity of nitrite and nitrosatable compounds administered orally to adult and prenatal animals; tumors were induced in many organs, including the stomach, esophagus, and nasal cavity (NAS 1981).

More than 120 *N*-nitroso compounds have been tested for carcinogenicity, and more than 75% of these compounds were shown to be carcinogenic (Shank and Magee 1981). These compounds were demonstrated to be carcinogenic in at least 22 species and carcinogenic transplacentally in at least five species (Schmahl and Habs 1980). All species tested have shown tumor formation after treatment with at least one of the *N*-nitroso compounds tested. Tumors have been induced in every organ and tissue and most cell types. Although organ specificity was observed within a species even after administration by different routes, clear differences in target tissue have been noted among species (Lijinsky et al. 1975).

IV. Quantification of Toxicological Effects[a]

The available data suggested that calculation of the HA values for nitrate/nitrite should include the following considerations: recognition of the newborn infant as the population group at greatest risk, recognition of the conversion of orally ingested nitrate to nitrite, and use of human data whenever possible because of the extreme quantitative and qualitative variations in nitrate and nitrite toxicity observed among species.

HA values are presented for a 4-kg infant (assumed to consume 0.64 L formula/d) and a 70-kg adult. HAs are usually determined for a 10-kg child and a 70-kg adult. However, newborn infants, assumed to weigh 4 kg, are the population subgroup at greatest risk, and thus HAs are provided for a 4-kg infant.

Although no separate HAs for a 10-kg child are provided, the HAs for a 70-kg adult are protective for all age groups other than a 4-kg infant, since they are based on data obtained in children (Craun et al. 1981).

Nitrate is toxic because it is converted to nitrite, and therefore the toxicity of nitrate and nitrite must be additive. Thus, nitrate and nitrite cannot be considered independently. Each HA is expressed as both mg nitrate-nitrogen/L drinking water and mg nitrite-nitrogen/L drinking water. Appropriate use of these values requires information on both the nitrate and nitrite content of drinking water so that a total "effective" nitrate concentration can be calculated and used as follows:

The "effective" nitrate-nitrogen concentration (mg/L) for all age groups is equal to nitrate-nitrogen + 10 × nitrite-nitrogen.

The "effective" nitrate-nitrogen concentration (mg/L) should not exceed the appropriate nitrate standard for the appropriate group (4-kg infant or 70-kg adult) or exposure period.

A. One-Day Health Advisory

The available data are insufficient to develop One-Day HAs for nitrate and nitrite. The Ten-Day HA should be protective of 1-d exposures.

B. Ten-Day Health Advisory

1. For Populations Other than 4-kg Infant. Craun et al. (1981) conducted an epidemiologic study of 102 children aged 1 to 8 yr in Washington County, Illinois. Of the study subjects, 64 consumed water with high nitrate levels (22 to 111 mg/L nitrate-nitrogen) and 38 consumed water with low nitrate levels (< 10 mg/L nitrate-nitrogen). Ingestion of water containing 22 to 111 mg/L nitrate-nitrogen did not result in abnormal mean methemoglobin levels and was not related to increased methemoglobin levels in comparison to controls.

[a] See Introduction for discussion of methodology

In the total study group of 102 children, only five had methemoglobin levels >2% (maximum of 3.1% in a child from the low-exposure group).

For a 70-kg adult and all age groups other than a 4-kg infant, the Ten-Day nitrate HA value is 111 mg/L nitrate-nitrogen, the NOAEL observed by Craun et al. (1981). Because the study was based on observations in humans and the most sensitive subgroup (i.e., infants) was considered separately, no uncertainty factor has been used in deriving the Ten-Day nitrate HA from the NOAEL.

No available studies provide a direct measure of the NOAEL for nitrite in children. The Ten-Day nitrite HA for a 70-kg adult and all other age groups other than a 4-kg infant can be calculated from the NOAEL for nitrate, assuming 10% conversion of nitrate to nitrite, as follows:

$$(111 \text{ mg/L nitrate-nitrogen})\ (0.10) = 11 \text{ mg/L nitrite-nitrogen}$$

where:

111 mg/L = NOAEL for nitrate based on the absence of methemoglobinemia in children; and 0.10 = assumed 10% conversion of nitrate to nitrite by a 10-kg child.

2. For a 4-kg Infant. A survey by the American Public Health Association (Walton 1951) found more than 278 cases of cyanosis in infants that were definitely associated with consumption of nitrate-contaminated water by the infant or nursing mother. No cases associated with water containing 10 mg/L or less of nitrate-nitrogen were found. As previously noted, Hegesh and Shiloah (1982) demonstrated that nitrites were synthesized in infants with acute diarrhea. Nitrites are responsible for methemoglobinemia, and thus it is possible that infants with diarrhea may be the population most sensitive to the toxic effects of both nitrate and nitrite. Because diarrhea is relatively common in infants, at least some of the infants noted in Walton (1951) were believed to have had diarrhea (USEPA 1985). Thus, it was concluded that a Ten-Day HA derived from the study by Walton (1951) could protect all infants, including those with diarrhea.

Based on the previous discussion, the Ten-Day nitrate HA for 4-kg infants is 10 mg/L nitrate-nitrogen, which is also the NOAEL for methemoglobinemia observed by Walton (1951). Studies by Donahoe (1949), Winton et al. (1971), and Toussaint and Wurkert (1982) support this HA.

No study provides a direct measure of the NOAEL for nitrite in infants. However, the Ten-Day nitrite HA for a 4-kg infant can be calculated from the NOAEL for nitrate as follows:

$$\frac{(10 \text{ mg/L nitrate-nitrogen})\ (100\%)}{(10)} = 1 \text{ mg/L nitrite-nitrogen } (1{,}000\ \mu\text{g/L})$$

where 10 mg/L = NOAEL for nitrate-nitrogen based on the absence of methomoglobinemia in infants; 100% = the assumed conversion of nitrate to nitrite by a 4-kg infant; and 10 = uncertainty factor.

C. Longer-Term Health Advisory

The available data are insufficient to develop Longer-Term HAs for nitrate and nitrite. However, for both nitrate and nitrite, the Ten-Day HA for a 4-kg infant is judged to offer protection against the formation of methemoglobin induced by the ingestion of either nitrate or nitrite in all age groups.

D. Lifetime Health Advisory

No suitable studies for calculation of a Lifetime HA were located. However, for both nitrate and nitrite, the Ten-Day HA for the 4-kg infant (10 mg/L nitrate-nitrogen and 1 mg/L nitrite-nitrogen) is judged to offer protection against the formation of methemoglobin induced by the ingestion of either nitrate or nitrite in all age groups.

As previously discussed, a 4-kg infant is the most sensitive member of the population with respect to the formation of methemoglobin induced either by nitrite directly or by in vivo reduction of nitrate to nitrite. In addition, as a 4-kg infant ages (e.g., becomes a 10-kg child), both sensitivity to the effects of methemoglobin and the amount of nitrate reduced to nitrite decrease, thus rendering the older child and the adult less sensitive to the effects of both nitrate and nitrite. Thus, the Ten-Day HA for a 4-kg infant for both nitrate and nitrite (10 mg/L nitrate-nitrogen and 1 mg/L nitrite-nitrogen) was concluded to offer adequate protection against methemoglobin formation in all other age groups as well.

E. Evaluation of Carcinogenic Potential

Animal studies provided inconclusive evidence regarding the carcinogenicity of nitrate and nitrite administered in the absence of nitrosatable compounds (USEPA 1985).

Applying the criteria described in USEPA guidelines for assessment of carcinogenic risk (USEPA 1986), both nitrate and nitrite may be classified in Group D: not classified. This category is for agents with inadequate or no animal evidence of carcinogenicity.

V. Other Criteria, Guidance, and Standards

The interim Maximum Contaminant Level (MCL) for nitrate-nitrogen is 10 mg/L (USEPA 1976b). The U.S. Public Health Service recommends a limit of 10 mg/L nitrate-nitrogen or 45 mg/L nitrate ion (USPHS 1962). The Committee on Water Quality Criteria of the National Academy of Sciences (NAS) recommends that nitrate-nitrogen concentration in public water supplies not exceed 10 mg/L and nitrite-nitrogen not exceed 1 mg/L (NAS 1972). The USEPA Quality Criteria for Water (USEPA 1976a) suggested the maximum concentrations of nitrate-nitrogen and nitrite-nitrogen in domestic water supplies not exceed 10 mg/L and 1 mg/L, respectively.

VI. Analytical Methods

Determination of nitrite alone or nitrite and nitrate combined is by colorimetry or spectrophotometry (USEPA 1979a,b). In these methods, a sample is passed through a column containing granulated copper-cadmium to reduce nitrate to nitrite. The nitrite (that which was originally present plus reduced nitrate) is determined by diazotizing with sulfanilamide and coupling with *N*-(1-naphthyl)-ethylenediamine dihydrochloride to form a highly colored azo dye, which then is measured colorimetrically or spectrophotometrically. Quantification of nitrate and nitrite separately is accomplished by carrying out the procedure first with and then without the copper-cadmium reduction step. The applicable range of the colorimetric and spectrophotometric methods is 0.05 to 10 mg/L nitrate-nitrogen and 0.01 to 1 mg/L nitrite-nitrogen, respectively.

An alternative reduction procedure may be used (USEPA 1979c). In this method, nitrate is reduced to nitrite with hydrazine sulfate. The applicable range of this method is 0.01 to 10 mg/L.

VII. Treatment Technologies

Ion exchange and reverse osmosis are the practical methods currently used to remove nitrates from water. Conventional coagulation and lime softening are not effective treatment methods for the removal of this contaminant (Laverentz 1974; USEPA 1977).

The application of ion-exchange resins for nitrate removal has a well-established history and is recognized as a practical treatment for drinking water systems (Gillies 1978; Sorg 1978; Sorg 1980; USEPA 1977).

Laboratory experiments and pilot plant studies showed that some strong base and weak base ion-exchange resins are nitrate selective and can reduce the nitrate concentration from as high as 50 mg/L (as N) to 0.5 mg/L (Gaundett 1975; Gregg 1973; Holzmacher 1971; Korngold 1973; Keulow et al. 1975). One full-scale ion-exchange plant has been operating successfully on Long Island, New York, since 1974. This plant lowers the nitrate level of 20 to 30 mg/L in the raw water to 0.5 mg/L. The finished water is a blend of treated and raw water and contains about 5 mg/L of nitrate (as N). Other installations removing nitrate include a 40,000 gal (151,400 L)/d plant at Curryville, Pennsylvania, and a 2,500 gal (9,463 L)/d plant in the Virgin Islands.

An important feature of the commercial nitrate ion-exchange resin is that the sulfate ion, not the nitrate, is the most preferred ion in the exchange.

However, field studies by Guter (1982) in McFarland, California, showed that nitrates can be removed effectively in the presence of sulfates as high as 380 mg/L.

Although reverse-osmosis systems have not been installed specifically to remove nitrates, removal efficiencies of 67 to 95% (high pressure) have been demonstrated. Two plants currently operating can provide data on nitrate removal. Laverentz (1974) reported that in Greenfield, Iowa, nitrate is reduced from 0.2 mg/L NO_3-N to 0.014 ng/L NO_3-N. In San Diego County Estates, Ramona, California, nitrate is reduced from 12.4 mg/L NO_3-N to 4.2 mg/L NO_3-N. However, considerable experimental field data were obtained when cellulose acetate was the only commercial membrane, and more recent field tests indicated nitrate rejection ranges for cellulose acetate membranes from 70 to 80% (Sourirajan 1977), 80 to 90% (Metcalf and Eddy 1979), and 58 to 86% (Weber 1972).

References

Archer MC, Tannenbaum SR, Fan T-Y, Weisman M (1975) Reaction of nitrite with ascorbate and its relation to nitrosamine formation. J Natl Cancer Inst 54:1203–1205.

Bloomfield RA, Hersey JR, Welsch CW, Garner GB, Muhrer ME (1962) Gastric concentration of nitrate in rats. J An Sci 21:1019.

Burden EHWJ (1961) The toxicology of nitrates and nitrites with particular reference to the potability of water supplies. Analyst 86:429–433.

Couch DB, Friedman MA (1975) Interactive mutagenicity of sodium nitrite, dimethylamine, methylurea and ethylurea. Mutat Res 31:109–114.

Craun GF, Greathouse DG, Gunderson DH (1981) Methemoglobin levels in young children consuming high nitrate well water in the United States. Int J Epidemiol 10:309–317.

Davidson KL, Hansel W, Crook L, McEntee K, Wright MJ (1964) Nitrate toxicity in dairy heifers. I. Effects on reproduction, growth, lactation and vitamin A nutrition. J Dairy Sci 47:1065–1073.

Druckery H, Steinhoff D, Beuthner H, Schneider H, Klarner P (1963) Screening of nitrate for chronic toxicity in rats. Arzneim Forsch 13:320–323. (In German; summary in English.)

Fritsch P, Canal M, Saint-Blanquat G, Hollande E (1980) Nutritional and toxicological impacts of nitrates and nitrites chronically administered (6 months) in rats. Ann Nutr Alim 34:1097–1114.

Gaundett RB (1975) Nitrate removal from water by ion exchange. Water Treat Exam 24(3):172–190.

Gillies MT (1978) Drinking water detoxification. Noyes Data Corporation.

Globus M, Samuel D (1978) Effect of maternally administered sodium nitrite on hepatic erythropoiesis in fetal CD-1 mice. Teratology 18:367–377.

Gregg JC (1973) Nitrate removal at water treatment plant. Civ Eng 43(4):45–47.

Gruener N, Shuval HI, Behroozi K, Cohen S, Shechter H (1973) Methemoglobinemia induced by transplacental passage of nitrites in rats. Bull Environ Contam Toxicol 9:44–48.

Guter GA[a] (1982) Removal of nitrate from contaminated water supplies for public use. Final report. USEPA Report no. EPA-600/82-042. USEPA Office of Drinking Water, Washington, DC.

Hegesh E, Shiloah J (1982) Blood nitrates and infantile methemoglobinemia. Clinica Chimica Acta 125: 107–115.

Holzmacher RG (1971) Nitrate removal from a ground water supply. Water Sewage Works 118(7): 210–213.

Hussain S, Ehrenberg L (1974) Mutagenicity of primary amines combined with nitrite. Mutat Res 26: 419–422.

Imaizumi S, Tyuma I, Imai K, Kosaka H, Ueda Y (1980) In vivo studies on methemoglobin formation by sodium nitrite. Int Arch Occup Environ Hlth 45: 97–104.

Inai K, Aoki Y, Tokuoka S (1979) Chronic toxicity of sodium nitrite in mice, with reference to its tumorigenicity. Gann 70: 203–208.

Ivankovic S (1979) Teratogenic and carcinogenic effects of some chemicals during prenatal life in rats, Syrian golden hamsters, and guinea pigs. Natl Cancer Inst Monogr 51: 103–115.

Kamm JJ, Dashman T, Newmark H, Mergens WJ (1977) Inhibition of amine-nitrite hepatotoxicity by alpha-tocopherol. Toxicol Appl Pharmacol 41: 575–583.

Keulow RW, Kropp KL, Withered J, Symons JM (1975) Nitrate removal by anion-exchange resins. J Am Waterworks Assoc 67(9): 528–534.

Kodama F, Umeda M, Tsutsui T (1976) Mutagenic effect of sodium nitrite on cultured mouse cells. Mutat Res 40: 119–124.

Korngold E (1973) Removal of nitrates from potable water by ion exchange. Water, Air, Soil Pollut 2: 15–22.

Laverentz DL (1974) Economic feasibility of desalting systems for municipal water supply in Iowa. U.S. Department of the Interior, Washington, DC.

Lijinsky W, Singer GM, Taylor HW (1975) Carcinogenic *N*-nitroso compounds. Proc XI Int Cancer Congress. 3: 44–47.

Maekawa A, Ogiu T, Onoderatt H, Furatak K, Matsuoka C, Ohno Y, Odashima S (1982) Carcinogenicity studies of sodium nitrite and sodium nitrate in F344 rats. Food Chem Toxicol 20: 25–33.

Metcalf and Eddy, Inc (1979) Wastewater engineering: Treatment, disposal, reuse, 2nd Ed. McGraw-Hill Co, New York, NY.

NAS (1972) National Academy of Sciences. Water quality criteria. EPA R3-73-033. National Academy Press, Washington, DC.

NAS (1978) National Academy of Sciences. Nitrates: an environmental assessment. National Academy Press, Washington, DC.

NAS (1981) National Academy of Sciences. The health effects of nitrate, nitrite and *N*-nitroso compounds. National Academy Press, Washington, DC.

Newberne PM (1978) Dietary nitrite in the rat. Final report on contract FDA-74-2181, Food and Drug Administration (FDA), Public Health Service (PHS), U.S. Department of Health, Education and Welfare (DHEW), Rockville, MD.

---

[a]USEPA documents can be obtained by writing to the appropriate office within the U.S. Environmental Protection Agency, 401 M Street, S W , Washington, DC 20460 They may also be available by contacting the National Technical Information Service, 5285 Port Royal Road, Springfield, VA 22161 (Telephone 1-800-336-4700).

Newberne PM (1979) Nitrite promotes lymphoma incidence in rats. Science 204:1079–1081.

Ohshima H, Bartsch H (1981) Quantitative estimation of endogenous nitrosation in humans by monitoring *N*-nitrosoproline excreted in the urine. Cancer Res 41: 3658–3662.

Parks NJ, Krohn KA, Mathis CA, Chasko JH, Geiger KR, Gregor ME, Peek NF (1981) Nitrogen-13-labeled nitrite and nitrate: Distribution and metabolism after intratracheal administration. Science 212:58–61.

Sander J, Schweinsberg F (1972) Interrelationships between nitrate, nitrite and carcinogenic *N*-nitroso compounds. 1. Communication: nitrate, nitrite and nitrosable amino-compounds in food and drugs, chemistry of *N*-nitroso compounds. Zentralbl Bakteriol Parasitenkd Infektionsk Hyg Abt 1: Orig Reihe B 156:299–340. (In German; summary in English.)

Schmahl D, Habs M (1980) Carcinogenicity of *N*-nitroso compounds. Species and route differences in regard to organotropism. Oncology 37:237–242.

Schneider NR, Yeary RA (1975) Nitrite and nitrate pharmacokinetics in the dog, sheep, and pony. Am J Vet Res 36.941–947.

Shank RC, Magee PN (1981) Toxicity and carcinogenicity of *N*-nitroso compounds. In: Shank RC (ed) Mycotoxins and *N*-nitroso compounds: environmental risks, Vol I. CRC Press, Boca Raton, FL. pp 185–217.

Shuval HI, Gruener N (1977) Health effects of nitrates in water. Report no. EPA-600/1-77-030[a]. USEPA Health Effects Research Laboratory, Cincinnati, OH.

Skrivan J (1971) Methemoglobin in pregnancy. Acta Univ Carol Med 17:123–161.

Smith JE, Beutler E (1966) Methemoglobin formation and reduction in man and various animal species. Am J Physiol 210:347–350.

Sorg TJ (1978) Treatment technology to meet the interim primary drinking water regulations for inorganics. J Am Waterworks Assoc 70(2):105–112.

Sorg TJ (1980) Compare nitrate removal methods. Water and Wastes Eng. 17(12):26–31.

Sourirajan S (1977) Reverse osmosis and synthetic membranes. National Research Council Canada. NRCC no. 15627. Ottawa, Canada.

Spiegelhalder B, Eisenbrand G, Preussmann R (1976) Influence of dietary nitrate on nitrite content of human saliva: possible relevance to in vivo formation of *N*-nitroso compounds. Food Chem Toxicol 14:545–548.

Sugiyami K, Tanaka T, Mori H (1979) Carcinogenicity examination of sodium nitrate in mice. Gifu Daigaku Igakubu Koyo. 27:1–6. (In Japanese; summary in English.)

Teramoto S, Saito R, Shirasu Y (1980) Teratogenic effects of combined administration of ethylenethiourea and nitrite in mice. Teratology 21:71–78.

Toussaint VW, Wurkert K (1982) Methamoglobinamie im Sauglingsalter. In: Selenka F (ed) Nitrat-Nitrit-Nitrosamine in Gewassern. Bonn, Germany: Deutsche Forschungsgemeinschaft, pp 136–142.

USEPA[a] (1976a) Quality criteria for water EPA 440-9-76-023. USEPA Office of Water Planning and Standards, Washington, DC.

USEPA[a] (1976b) National interim primary drinking water regulations. Report no. EPA-570/9-76-003 USEPA Office of Drinking Water, Washington, DC.

USEPA[a] (1977) Manual of treatment techniques for meeting the interim primary drinking water regulations, revised Report no. EPA-600/8-77-005. USEPA Office of Drinking Water, Washington, DC.

USEPA[a] (1979a) Method 353.2. Colorimetric, automated, cadmium reduction. In: Methods for chemical analysis of water and wastes. Report no. EPA-600/4-79-020. March. USEPA Environmental Monitoring and Support Laboratory, Cincinnati, OH.

USEPA[a] (1979b) Method 353.3. Spectrophotometric, cadmium reduction. In: Methods for chemical analysis of water and wastes. Report no. EPA-600/4-79-020. March. USEPA Environmental Monitoring and Support Laboratory, Cincinnati, OH.

USEPA[a] (1979c) Method 353.1. Colorimetric, automated, hydrazine reduction. Methods for chemical analysis of water and wastes. Report no. EPA-600/4-79-020. March. USEPA Environmental Monitoring and Support Laboratory, Cincinnati, OH.

USEPA[a] (1985) Health effects criteria document for nitrate/nitrite. USEPA Criteria and Standards Division, Office of Drinking Water, Washington, DC.

USEPA (1986) Guidelines for carcinogenic risk assessment. Fed Regis 51(185): 33992–34003. September 24.

USEPA[a] (1987) Estimated national occurrence and exposure to nitrate and nitrite in public drinking water supplies. USEPA Criteria and Standards Division, Office of Drinking Water, Washington, DC.

USFDA (1972a) Food and Drug Administration Study of mutagenic effects of sodium nitrate (71-7). Contract FDA 71-267. Report no. FDABF-GRAS-083. Stanford Research Institute, Menlo Park, CA. 103 pp.

USFDA (1972b) Food and Drug Administration Study of mutagenic effects of sodium nitrate (71-9). Contract FDA 71-267. Report no. FDABF-GRAS-084. Stanford Research Institute, Menlo Park, CA. 103 pp.

USPHS (1962) U S. Public Health Service. U.S. Public Health Service drinking water standards. U.S Department of Health, Education and Welfare (DHEW), Rockville, MD.

Walton G (1951) Survey of literature relating to infant methemoglobinemia due to nitrate contaminated water. Am J Publ Hlth 41:986–996.

Weast RC ed (1974) The CRC handbook of chemistry and physics. CRC Press, Inc, Boca Raton, FL.

Weber WJ (1972) Physicochemical processes for water quality control. Wiley-Interscience, New York, NY.

WHO (1962) World Health Organization. Evaluation of the toxicity of a number of antimicrobials and antioxidants. Sixth report of the Joint FAO/WHO Expert Committee on Food Additives, World Health Organization Technical Report Series no. 228. WHO, Geneva, Switzerland.

Whong WZ, Speciner ND, Edward GS (1979) Mutagenicity detection of *in vivo* nitrosation of dimethylamine by nitrite. Environ Mutag 1:277–282.

Windholz M (ed) (1976) The Merck index—an encyclopedia of chemicals and drugs. 9th Ed, Merck & Co Inc, Rahway, NJ.

Winton EF, Tardiff RG, McCabe LJ (1971) Nitrate in drinking water. J Am Waterworks Assoc 63:95–98.

Witter JP, Gatley SJ, Balish E (1979) Distribution of nitrogen-13 from labeled nitrate ($^{13}NO_3^-$) in humans and rats. Science 204:411–413.

Styrene[a]

I. General Information

Styrene, also known as vinylbenzene, cinnamene, phenylethylene, and ethenylbenzene, is used in plastics. Its structure and properties are shown in Table 1. Styrene is produced primarily from dehydrogenation of ethylbenzene. In 1982, U.S. production of styrene totaled 5.9 billion lb (2.7 billion kg).

National drinking water surveys indicated styrene is an infrequent contaminant. To date, the testing of 941 ground-water supplies and 102 surface-water supplies has failed to result in detection of a single positive occurrence (Boland 1981).

Contamination of drinking water by styrene, however, has been reported occasionally by state programs.

II. Pharmacokinetics

A. Absorption

Available data indicate the absorption of styrene from the gastrointestinal tract of rats is rapid and virtually complete (Plotnick and Weigel 1979).

Styrene uptake and absorption have been the subject of a number of human inhalation studies (Fiserova-Bergerova and Teisinger 1965; Teramoto and Horiguchi 1979). The findings of these studies indicate that pulmonary retention of styrene is approximately two-thirds of the administered concentration with considerable variation in measured uptake between individuals and studies. Mean uptakes ranged from 59 to 89%.

B. Distribution

The distribution of styrene after oral administration was studied in rats given single doses of 20 mg/kg [^{14}C]-styrene in corn oil by gavage (Plotnick and

[a] Based on USEPA (1988)

Reviews of Environmental Contamination and Toxicology, Vol 107

Table 1. General information and properties of styrene

Property	Value
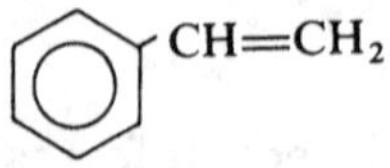	
Chemical Abstracts Service (CAS) number	100-42-5
Chemical formula	C_8H_8
Molecular weight	104.16
Physical state	Clear, colorless liquid with characteristic sweet, pleasant odor
Boiling point	145°C
Melting point	−30.6°C
Density (20°C)	0.9 g/mL
Vapor pressure (20°C)	4 5 mm Hg
Specific gravity	0.90
Water solubility	320 mg/L
Log octanol water/partition coefficient	2.95
Odor threshold	0 01 mg/L (water)
Taste threshold	—
Air concentration conversion factor	4.26 mg/m^3 = 1 ppm

Source Hansch and Leo (1979), Lewis et al (1983)

Weigel 1979). Peak tissue levels were reached within 2 to 4 hr. The organs with the highest concentrations were the kidney (46 μg/g in males, 25 μg/g in females), liver (13 μg/g in males, 7 μg/g in females), and pancreas (10 μg/g in males, 6 μg/g in females), with lower concentrations observed in lungs, heart, spleen, adrenals, brain, testes, and ovaries.

Results of inhalation studies in rats indicate that distribution of styrene is widespread with relatively high concentrations accumulating in adipose tissue (Withey and Collins 1979).

In humans, Dowty et al. (1976) found concentrations of transplacentally transferred styrene to be somewhat higher than those of maternal blood, which suggests a selective one-way transplacental transfer.

Pellizzari et al. (1982) detected styrene in each of eight milk samples collected from lactating women residing in various cities.

C. Metabolism

The metabolic fate of styrene in mammals has been studied extensively. Information from human studies is limited, but similarities to the process in other mammals have been identified.

Based on studies in rats administered styrene-7,8-oxide or styrene glycol by intraperitoneal injection, Ohtsuji and Ikeda (1971) proposed that the metabolism of styrene proceeds via P-450 microsomal oxidation to styrene oxide, styrene glycol, and then to mandelic acid, which is metabolized to phenylglyoxylic acid or hippuric (urobenzoic) acid.

D. Excretion

Results of a number of studies in rats (Ramsey and Young 1978, 1980; Teramoto and Horiguchi 1979; Withey and Collins 1977, 1979) indicated that styrene is eliminated relatively rapidly from all tissues in test animals.

Twenty-four hours after oral administration of 20 mg/kg [^{14}C]-styrene to rats, concentrations in all tissues and organs examined were $< 1\ \mu g/g$ (Plotnick and Weigel 1979).

Elimination of styrene from the heart, brain, liver, spleen, and kidneys of rats was described by biphasic log-linear kinetics after IV injection of 4.0 mg/kg (Withey and Collins 1977). Half-lives ranged from 3.8 to 7.1 min for the alpha (fast) phase and from 20 to 37 min for the beta (slow) phase.

Predictions based on a toxicokinetic model (parameters estimated from a human inhalation study) indicate that maximum concentrations of styrene in both blood and fat of humans are reached after a few repeated 8-hr daily exposures to 80 ppm styrene (Ramsey et al. 1980; Ramsey and Young 1978, 1980). These data suggest no tendency toward long-term accumulation exists.

III. Health Effects

A. Humans

Results of controlled experiments using human volunteers indicated styrene administered by inhalation at relatively high doses results in central nervous system (CNS) effects.

Drowsiness, listlessness, and an altered sense of balance were reported during a 4-hr exposure of two male subjects to styrene at 3,407 mg/m^3 (800 ppm) (Carpenter et al. 1944).

Stewart et al. (1968) reported that volunteers exposed to styrene by inhalation at 217 mg/m^3 (50 ppm) and 499 mg/m^3 (117 ppm) for 1 and 2 hr, respectively, showed no signs of toxicity. The moderately strong initial styrene odor diminished after 5 min. At 921 mg/m^3 (216 ppm), nasal irritation resulted after 20 min. Eye and nose irritation, strong odor, and altered neurologic function were reported for volunteers exposed to styrene at 1,600 mg/m^3 (376 ppm) for 1 hr. Most volunteers exposed to this level exhibited reduced performance in the Crawford manual dexterity collar-and-pin test, the modified Romberg test, and the Flannagan coordination test. Six subjects were exposed to 422 mg/m^3 (99 ppm) styrene vapor for 7 hr. No serious untoward effects were noted at this concentration.

Gamberale and Hultengren (1974) exposed 12 subjects to styrene by inhalation at concentrations of 213, 639, 1,065, and 1,491 mg/m^3 (50, 150, 250, and 350 ppm) during four consecutive 30-min intervals. A dose-related increase in single reaction time was evident. Reaction time recorded during the final 30-min exposure was significantly increased ($p < 0.05$).

Odkvist et al. (1982) studied the effects of styrene on vestibulo-oculomotor functions in 10 subjects exposed to styrene by inhalation at 370 to 591 mg/m^3 (88 to 140 ppm) for approximately 80 min. The rate of saccadic movement (of the eyes between two alternating light sources increased significantly ($p < 0.05$) after exposure. Suppression of the vestibulo-oculomotor reflex was also affected.

Evidence suggests the human fetus is more sensitive than the adult to the toxic effects of styrene (Holmberg 1977; Hemminki et al. 1980).

The frequency of spontaneous abortions among Finnish chemical workers was analyzed by Hemminki et al. (1980). Information on spontaneous abortions (15,482 cases), induced abortions (71,235 cases), and births (193,897 cases) for 1973 to 1976 was obtained from the Hospital Discharge Registry of the Finnish National Board of Health. Cases were matched by social security number to the membership of the Finnish Union of Chemical Workers (approximately 900 female members). About 85% of the total number of spontaneous abortions in Finland were reportedly listed in the Hospital Discharge Registry. The rate of spontaneous abortions was defined as the number of spontaneous abortions × 100/number of pregnancies. The rates of spontaneous abortion were 8.54% ($N = 52$) and 15.0% ($N = 6$) among the female union members and a subgroup in the styrene industry, respectively. These rates were significantly higher ($p < 0.01$) than the rate among all Finnish women (5.52%; 15,482 spontaneous abortions). The rates of spontaneous abortion were 15.57 and 31.59% in the female union workers and female styrene industry workers, respectively. These are significantly higher ($p < 0.001$) than the rate among all Finnish women (7.98%).

The information on the work histories of 43 Finnish mothers of children with central nervous system (CNS) defects born from June 1, 1976 to March 1, 1977 was obtained through personal interviews (Holmberg 1977). Two of these mothers had been employed in the reinforced plastics industry and had regular exposure to styrene, polyester resin, organic peroxides, and acetone during pregnancy. The defects in their two children were anencephaly and congenital hydrocephaly. The overall rates of anencephaly and congenital hydrocephaly were reported to be 0.2 and 0.3, respectively, per 1,000 live births in Finland. Based on these estimates, the rate of these malformations in the offspring of mothers in the reinforced plastics industry appeared to be 300-fold higher than the rate in the general population (2 of 12 vs 0.5 of 1,000) during the 9-mon study.

B. Animals

1. Short-Term Exposure. Wolf et al. (1956) reported an acute oral LD_{50} of greater than 5,000 mg/kg for rats treated with styrene by gavage. This indicates that the acute toxicity of styrene was relatively low.

The lowest single oral dose of styrene (administered by oral intubation) causing 100% mortality in rats within 2 wk of treatment was 8,000 mg/kg; 1,600 mg/kg was the maximum dose resulting in no deaths (Spencer et al. 1942).

The effects of styrene administration at 270, 450, or 900 mg/kg orally (method not stated) for 7 consecutive d on hepatic mixed-function oxidase enzyme activities, glutathione content, and glutathione-*S*-transferase activity were reported by Das et al. (1981). Activities of aryl-hydrocarbon hydroxylase and aniline hydroxylase were significantly enhanced at higher doses of styrene (450 and 900 mg/kg). A significant decrease in glutathione content and inhibition of glutathione-*S*-transferase activity were also noted at the highest dose of styrene (900 mg/kg). Therefore, the No-Observed-Adverse-Effect Level (NOAEL) for effects on hepatic enzymes in this study was 270 mg/kg/d.

Agrawal et al. (1982) studied the effects of styrene on dopamine-receptor binding in rats. Styrene was administered at 200 or 400 mg/kg/d by gavage to groups of six 8-wk-old ITRC male albino rats. Styrene was administered in a single dose or in up to 90 daily doses over 90 d. Significant increases in the specific binding of [^{3}H]-spiroperidol to dopamine receptors in the corpus striatum were noted at both levels after single or repeated exposure to styrene. The Lowest-Observed-Adverse-Effect Level (LOAEL) for this study was identified as 200 mg/kg/d.

2. Long-Term Exposure. Changes in hepatic enzyme activity after oral exposure to styrene have been demonstrated by a number of investigators.

Srivastava et al. (1982) administered styrene by gavage (at 200 or 400 mg/kg/d) to groups of five adult male albino ITRC rats, 6 d/wk, for 100 d. These animals exhibited no changes in weight gain and no other overt signs of toxicity. Significant dose-dependent increases in hepatic enzymes (benzo[*a*]pyrene hydroxylase and aminopyrine-*N*-demethylase) as well as decreases in glutathione-*S*-transferase were observed. Some mitochondrial enzymes were significantly decreased as well. Histopathologic changes, seen at only the high dose, involved tiny areas of focal liver necrosis, consisting of a few degenerated hepatocytes and inflammatory cells. Therefore, the LOAEL for hepatic effects in this study was 200 mg/kg/d.

Groups of 10 female rats were administered styrene at 66.7, 133, 400, or 667 mg/kg/d by intubation, 5 d/wk, for 6 mon (Wolf et al. 1956). At the two higher dose levels, decreased growth weights and increased liver and kidney weights were observed without hematologic or histopathologic effects. At the two lower dose levels, no effects were noted in body weight, organ weight, or pathologic findings. Therefore, the NOAEL for this study was 133 mg/kg/d and the LOAEL was 400 mg/kg/d.

Beagle dogs were given styrene in a peanut oil suspension by gavage 7 d/wk for 561 d (Quast et al. 1978, 1979). Dose levels were 200, 400, or 600 mg/kg/d. The controls received peanut oil only. At the two higher dose levels, minimal histopathologic effects were noted in the liver (i.e., increased iron deposits within the reticuloendothelial cells). Hematologic effects included increased Heinz bodies in erythrocytes and a decreased packed-cell volume. At the lowest dose level, these effects were not observed. Therefore, 200 mg/kg/d was identified as the NOAEL for this study and 400 mg/kg/d can be designated as the LOAEL.

3. Reproductive Effects. The reproductive and teratogenic effects of styrene oxide were assessed in Wistar rats (Sikov et al. 1981). The percentage of pregnant rats was significantly reduced.

4. Developmental Effects. Investigators at Dow Chemical Company administered styrene in peanut oil to pregnant Sprague–Dawley rats (29 to 39 dams/group) by gavage at dose levels of 0, 180, or 300 mg/kg/d (0, 90, 150 mg/kg twice daily) on d 6 through 15 of gestation (Murray et al. 1976, 1978). Maternal toxicity was indicated by significantly reduced ($p < 0.05$) body weight gain and food consumption at both dose levels during d 6 to 9 of gestation. No significant effects on maternal mortality or percentage of pregnancies were observed. No developmental effects were observed.

5. Mutagenicity. Results were negative for six mutagenicity tests using *Salmonella typhimurium* test systems, with and without S-9 metabolic activation. Styrene was tested using strains TA1535, TA1537, TA98, and TA100. De Meester et al. (1977, 1981) and Vainio et al. (1976) obtained positive results with mutant strains sensitive to base-pair substitution, and all test results were negative in strains sensitive to frame-shift mutagens.

Styrene oxide, a major metabolite of styrene, has been demonstrated consistently to be mutagenic in *S. typhimurium* TA1535 and TA100, in the presence and absence of a mammalian metabolic activating system (De Meester et al. 1977, 1981).

6. Carcinogenicity. Several carcinogenicity studies have reported statistically significant increases in the incidences of tumors at specific sites in specific test groups following styrene exposure. Increased tumor incidences were observed in three studies (Jersey et al. 1978; NCI 1979; Ponomarkov and Tomatis 1978). Some increases in tumor incidences were statistically significant when compared with concurrent controls.

Direct evidence that inhalation of styrene causes cancer in laboratory animals comes from the study by Jersey et al. (1978). Groups of male or female Sprague-Dawley rats were exposed for 6 h/d, 5 d/wk to styrene vapor at 600 or 1,200 ppm. The high dose was reduced to 1,000 ppm after 2 mon. The study was terminated after 24 mon. A statistically significant increase in mammary adenocarcinoma was observed in females exposed to 600 ppm.

Evidence that ingestion of styrene causes cancer in animals comes from two studies. Ponomarkov and Tomatis (1978) administered styrene by olive oil gavage to female O_{20} mice, C57B1 mice, and BDIV rats once on d 17 of gestation and then weekly throughout their offspring's lifetimes. A statistically significant increased incidence and earlier onset of lung tumors were observed in the O_{20} offspring, but there was a high background tumor rate in this strain. In the C57B1 offspring there was a statistically significant increased incidence of liver tumors compared with pooled, untreated, and vehicle controls.

NCI (1979) administered styrene at 150 or 300 mg/kg daily by corn oil gavage to B6C3F1 mice. Exposure was terminated after 78 wk, and the study was terminated after 91 wk. Statistically significant increased incidences of lung alveolar/bronchiolar adenomas or carcinomas were observed in both exposed groups with a statistically significant dose-related trend.

Negative results were observed in a chronic bioassay investigating induced tumors of the brain (Maltoni et al. 1982). Sprague–Dawley rats were exposed to styrene in olive oil by gavage (0, 50, or 250 mg/kg) or inhalation to styrene vapors (0, 25, 50, 100, 200, or 300 ppm). Styrene was not found to induce brain tumors.

Styrene oxide, a metabolite of styrene, produced a dose-related increase in the incidence of stomach tumors in rats (oral doses of 4.12 or 8.25 mg/kg four to five times/wk for 52 wk) (Maltoni et al. 1979). Styrene oxide administered by gavage three times/wk for 2 yr to male and female F344 rats and B6C3F1 mice (low and high doses of 275 and 500 mg/kg for rats and 375 and 750 mg/kg for mice) produced high incidences of forestomach carcinomas and papillomas in all treatment groups. Further, the incidence of hepatocellular adenomas and carcinomas was significantly elevated in the low-dose male mice relative to controls (Lijinsky 1986). Styrene oxide (100 to 150 mg/kg) also produced statistically significant benign and malignant forestomach tumors in both sexes of BDIV rats after gavage administration of 96 wk. In addition, two other tumor types (lung and nervous system) rare to this strain were noted;

however, their incidences did not reach statistically significant levels (Ponomarkov et al. 1984).

Retrospective cohort mortality and case-control studies were conducted in workers exposed to styrene in the styrene-polystyrene manufacturing industry and in the styrene-butadiene synthetic rubber industry (McMichael et al. 1976; Meinhardt et al. 1978; Smith and Ellis 1977). Data are now inadequate to conclude that styrene is a human carcinogen. However, an elevated incidence of tumors of hematopoietic and lymphatic tissues has been observed. The available studies are limited because of relatively small cohort sizes or multiple chemical exposures (including exposure to benzene).

IV. Quantification of Toxicological Effects[a]

A. One-Day Health Advisory

The study of human inhalation exposure to styrene by Stewart et al. (1968) was selected to serve as the basis for calculating the One-Day Health Advisory (HA). The study involved a controlled styrene inhalation exposure using nine healthy human male volunteers. No subjective or objective signs of toxicity were noted following 1- and 2-hr exposures to 51 ppm (217 mg/m^3) or 117 ppm (440 mg/m^3) styrene, respectively. To simulate a work day, six subjects were exposed to 99 ppm (422 mg/m^3) styrene vapor for 7 hr. No serious untoward effects were noted except mild eye and throat irritation in three subjects. There were no objective signs of impairment of balance or coordination; however, three of the six subjects did report that they had intermittent difficulty in performing the modified Romberg Test. In contrast, exposure to 376 ppm (1,602 mg/m^3) styrene vapor for 2 hr resulted in abnormal neurological findings and complaints of nausea and inebriation. The results of urinalysis and hematology and blood chemistry studies were normal and unchanged from pre-exposure values.

The results of another study (Odkvist et al. 1982) using human volunteers exposed to similar styrene levels indicate that the mean pulmonary styrene uptake is 64% of the inspired amount. Using a NOAEL of 99 ppm (422 mg/m^3) from a 7-hr exposure, the One-Day HA for a 10-kg child can be derived. First, the total absorbed dose (TAD) is determined.

$$\text{TAD} = \frac{(442\ \text{mg/m}^3)(20\ \text{m}^3/\text{d})(7\ \text{hr}/24\ \text{hr})(0.64)}{(70\ \text{kg})} = 22.5\ \text{mg/kg/d}$$

where:

7 hr/24 hr = duration of exposure;
20 m^3/d = assumed ventilation volume for 70-kg adult;
0.64 = estimated ratio of absorbed dose (Odkvist et al. 1982).

[a] See Introduction for discussion of methodology

Therefore, the One-Day HA for a 10-kg child, based on a TAD of 22.5 mg/kg/d, is:

$$\text{One-Day HA} = \frac{(22.5 \text{ mg/kg/d})(10 \text{ kg})}{(10)(1 \text{ L/d})} = 22.5 \text{ mg/L}$$

B. Ten-Day Health Advisory

No information was found in the available literature that was suitable for deriving a Ten-Day HA value for styrene. It is therefore recommended that the Longer-Term HA for a 10-kg child (2 mg/L) be used at this time as a conservative estimate of the Ten-Day HA.

C. Longer-Term Health Advisory

The Quast et al. (1978, 1979) gavage study in dogs was chosen to serve as the basis for calculating the Longer-Term HA. The study identified a NOAEL of 200 mg/kg/d based on the absence of hepatic and hematological effects. The Longer-Term HAs are calculated as follows:

For a 10-kg child:

$$\text{Longer-Term HA} = \frac{(200 \text{ mg/kg/d})(10 \text{ kg})}{(100)(10)(1 \text{ L/d})} = 2 \text{ mg/L } (2{,}000\ \mu\text{g/L})$$

For a 70-kg adult:

$$\text{Longer-Term HA} = \frac{(200 \text{ mg/kg/d})(70 \text{ kg})}{(100)(10)(2 \text{ L/d})} = 7 \text{ mg/L } (7{,}000\ \mu\text{g/L})$$

where 10 = modifying factor to reduce cancer risk.

D. Lifetime Health Advisory

For comparison with cancer-based criteria, the Drinking Water Equivalent Level (DWEL) may be calculated on the basis of the study in dogs by Quast et al. (1978). Using the NOAEL of 200 mg/kg/d based on the absence of hepatic and hematological effects, the DWEL is calculated as follows:

Step 1: Determination of the Reference Dose (RfD)

$$\text{RfD} = \frac{(200 \text{ mg/kg/d})}{(1{,}000)} = 0.2 \text{ mg/kg/d}$$

Step 2. Determination of the Drinking Water Equivalent Level (DWEL)

$$\text{DWEL} = \frac{(0.2 \text{ mg/kg/d})(70 \text{ kg})}{(2 \text{ L/d})} = 7 \text{ mg/L } (7{,}000\ \mu\text{g/L})$$

Step 3: Determination of Lifetime HA

A Lifetime HA is *not* recommended because styrene is currently classified in Group B2: probable human carcinogen.

The EPA is reevaluating its current position on the cancer classification of styrene. Should the previous classification of Group C be retained, calculation of a Lifetime HA would be recommended as follows:

Determination of the Lifetime Health Advisory

$$\text{Lifetime HA} = \frac{(7\text{ mg/L})(20\%)}{(10)} = 0.14\text{ mg/L (rounded to 100 }\mu\text{g/L)}$$

E. Evaluation of Carcinogenic Potential

The results of three chronic animal bioassays (Jersey et al. 1978; NCI 1979; Ponomarkov and Tomatis 1978) collectively provide sufficient animal evidence for the carcinogenic activity of styrene. Strong supporting evidence is provided by metabolic and genotoxicity studies. When the animal bioassay data are considered collectively along with the metabolism/genotoxicity data, there is a reasonable basis for classifying styrene as having a "sufficient" level of evidence. Therefore styrene is currently classified as a Group B2 chemical using EPA's Guidelines for Carcinogen Risk Assessment. The EPA Science Advisory Board (SAB) reviewed the Drinking Water Health Criteria Document for styrene (February 4–5, 1988) and provided written review comments (July 19, 1988). The extensive SAB review comments are currently being considered by the EPA. The SAB does not agree that there is sufficient evidence to justify the reclassification of styrene to EPA category B2 and recommends continuation of the Group C classification. After consideration of the SAB review and public comment, the EPA will reexamine this decision. EPA recognizes that the Group B2 classification is a matter for discussion due to different views on interpretation of the animal bioassay data. For example, the comparison of exposed animals to concurrent controls versus historical control data can result in different levels of statistical significance. Another critical issue affecting classification is the degree of emphasis to place on supporting metabolism and genotoxicity data.

Because the NCI study (NCI 1979) with male B6C3F1 mice included more than one exposure level and showed a dose-related trend, it is considered to provide a reliable basis for estimating the cancer potency factor ($q_1{}^*$). Using the linearized multistage model, the cancer potency factor is 3×10^{-2} $(\text{mg/kg/d})^{-1}$. The 10^{-4}, 10^{-5}, and 10^{-6} cancer risk levels are 100, 10, and 1 μg/L, respectively, for a lifetime exposure.

For comparison purposes, the 10^{-6} risk level for the multihit, one-hit, probit, logit, and Weibull models are <0.001, 0.5, 2.0, <0.001, and <0.001 μg/L, respectively.

The animal bioassay data on styrene-7,8-oxide were evaluated by IARC

(1985) and considered to provide sufficient evidence for carcinogenicity in experimental animals; no human data were available. On the basis of sufficient animal evidence and the absence of human data, IARC considered styrene-7,8-oxide to present a carcinogenic risk to humans (IARC 1985).

IARC has recently classified styrene in IARC group 2B (IARC 1987). The IARC group 2B includes chemicals for which there are inadequate or no data in humans, but limited evidence of carcinogenicity in experimental animals together with supporting evidence from other relevant data.

V. Other Criteria, Guidance, and Standards

The Occupational Safety and Health Administration (OSHA) standard for styrene is an 8-hr Time-Weighted Average (TWA) concentration of 100 ppm, a ceiling concentration of 200 ppm, and a maximum peak concentration (29 CFR 1910.1000; Table Z-2) of 600 ppm for 5 min or less in any 3-hr period.

The American Conference of Governmental Industrial Hygienists has established the TWA-TLV for styrene is workroom air as 50 ppm with a Short-Term Exposure Level (STEL) of 100 ppm (ACGIH 1982). The TLV was reduced from 100 ppm in 1981 (ACGIH 1981).

The National Institute for Occupational Safety and Health (NIOSH 1983) recommended a styrene concentration limit in workplace air of 50 ppm TWA for up to a 10-hr day, 40-hr work week and a ceiling concentration of 100 ppm determined during any 15-min sampling period.

VI. Analytical Methods

Styrene content is determined by a purge-and-trap gas chromatographic (GC) procedure used for the determination of volatile aromatic and unsaturated organic compounds in water (USEPA 1985a). This method calls for bubbling an inert gas through the sample and trapping styrene on an adsorbent. The adsorbent is heated to drive off styrene onto a GC column, which is temperature programmed to separate the method analytes, which are then detected by a photoionization detector. This method is applicable to the measurement of styrene over a concentration range of 0.05 to 1,500 μg/L. Confirmatory analysis for styrene is by mass spectrometry, which has a detection limit of 0.3 μg/L (USEPA 1985b).

VII. Treatment Technologies

Information is available on the removal of styrene from water by air stripping, adsorption, and oxidation. Styrene has a Henry's Law constant of 12 atm, which suggests it is amenable to removal from water by air stripping (USEPA 1985c).

Decarbonators that have some aeration function have been evaluated for their efficacy in styrene removal. When the influent styrene concentration was 0.076 μg/L, the decarbonators tested were able to remove 51.3% (USEPA 1985c).

Tests evaluating adsorption of styrene by granular-activated carbon showed that an average of 40% was removed over a 10-mon period (USEPA 1985c). The influent styrene concentration was 0.03 μg/L.

The ethenyl double bond of styrene renders it amenable to oxidation. It is, therefore, possible that oxidative techniques may be effective in removing styrene from potable water. Bench-scale evaluations of ozone treatment of styrene-contaminated water conducted by Avigne (1983, cited in USEPA 1985a) indicated the reaction rate constant for a 0.007 mM styrene solution (pH 2) is 300,000 L/mole-sec. The pH was maintained at 2 to inhibit the decomposition of ozone. Oxidation of styrene to benzaldehyde and hydrogen peroxide was reported by Legube (1983, cited in USEPA 1985a). Using an ozone application rate of 10^7 mg/hr at 12 L/hr, 0.9 moles ozone/mole styrene was required for complete oxidation of the styrene. The initial styrene concentration was 1.1×10^{-4} mole/L. It was suggested that further oxidation of benzaldehyde to benzoic acid might occur.

It is possible that other oxidizing agents, such as permanganate, could be effective in oxidizing styrene. However, no studies of tests of these alternative oxidizing agents are available.

References

ACGIH (1981) American Conference of Governmental Industrial Hygienists. TLVs —Threshold limit values for chemical substances and physical agents in the workroom environment with intended changes for 1981. ACGIH, Cincinnati, OH, p 50.

ACGIH (1982) American Conference of Governmental Industrial Hygienists. TLVs —Threshold limit values for chemical substances and physical agents in the workroom environment with intended changes for 1982. ACGIH, Cincinnati, OH, p 29.

Agrawal AK, Srivastava SP, Seth PK (1982) Effect of styrene on dopamine receptors. Bull Environ Contam Toxicol 29(4):400–403.

Boland PA[a] (1981) National screening program for organics in drinking water. EPA contract 68-01-4666. Stanford Research Institute (SRI) International. USEPA Office of Drinking Water, Washington, DC.

Carpenter CP, Shaffer CB, Weil CS, Smyth HF (1944) Studies on the inhalation of 1,3-butadiene with a comparison to its narcotic effect with benzol, toluol and styrene, and a note on the elimination of styrene by the human. J Ind Hyg Toxicol 26(3):69–78.

[a]USEPA documents can be obtained by writing to the appropriate office within the U S Environmental Protection Agency, 401 M Street, S W , Washington, DC 20460 They may also be available by contacting the National Technical Information Service, 5285 Port Royal Road, Springfield, VA 22161 (Telephone. 1-800-336-4700).

Das M, Dixit R, Mushtaq M, Srivastava SP, Seth PK (1981) Effect of styrene on hepatic mixed function oxidases, glutathione content and glutathione-*S*-transferase activity in rats. Drug Chem Toxicol 4(3):219–227.

De Meester C, Poncelet F, Roberfroid M, Rondelet J, Mercier M (1977) Mutagenicity of styrene and styrene oxide. Mutat Res 56(2):147–152.

De Meester C, Durverger-Van Bogaert M, Lambotte-Vandepaer M, Mercier M, Poncelet F (1981) Mutagenicity of styrene in the *Salmonella typhimurium* test system. Chem Biol Interact 20(2):163–170.

Dowty BJ, Laseter JL, Storer J (1976) Transplacental migration and accumulation in blood of volatile organic constituents. Pediatr Res 10:696–701.

Fiserova-Bergerova V, Teisinger J (1965) Pulmonary styrene vapor retention. Ind Med Surg 34:620–622.

Gamberale F, Hultengren M (1974) Exposure to styrene. II. Psychological functions. Work Environ Hlth 11(2):86–93.

Hansch C, Leo AJ (1979) Substituent constants for correlation analysis in chemistry and biology. John Wiley and Sons, New York, NY.

Hemminki K, Franssila E, Vainio H (1980) Spontaneous abortion among female chemical workers in Finland. Int Arch Occup Hlth 45:123–126.

Holmberg PC (1977) Central nervous defects in two children of mothers exposed to chemicals in the reinforced plastics industry. Scand J Work Environ Hlth 5: 333–335.

IARC (1987) International Agency for Research on Cancer. IARC monographs on the evaluation of the carcinogenic risk of chemicals to humans. Overall evaluations of carcinogenicity and updating of IARC monographs. Vol 1–42, suppl 7. IARC, Lyon, France, p 345–347.

IARC (1985) International Agency for Research on Cancer. IARC monographs on the evaluation of the carcinogenic risk of chemicals to humans: Alkyl compounds, aldehydes, epoxides, and peroxides. Vol 36. IARC, Lyon, France, pp 245–263.

Jersey G, Balmer M, Quast J, Park CN, Schuetz DJ, Beyer JE, Olson KJ, McCollister SB, Rampy LW (1978) Two-year chronic inhalation toxicity and carcinogenicity study on monomeric styrene in rats. Dow Chemical Study for Manufacturing Chemical Association. December 6.

Lewis PJ, Hagopian C, Koch P (1983) Styrene. In: Grayson M and Eckroth D (eds) Kirk-Othmer encyclopedia of chemical toxicology, Vol 21. 3rd Ed. John Wiley and Sons, Inc. New York, NY. pp 770–801.

Lijinsky W (1986) Rat and mouse forestomach tumors induced by chronic oral administration of styrene. J Natl Cancer Inst 77(2):471–476.

Maltoni C, Failla G, Kassapidis G (1979) First experimental demonstration of the carcinogenic effects of styrene oxide: Long-term bioassays on Sprague-Dawley rats by oral administration. Med Lav 70(50):358–362.

Maltoni C, Ciliberti A, Carretti D (1982) Experimental contributions in identifying brain potential carcinogens in the petrochemical industry. Ann New York Acad Sci 381:216–249.

McMichael AJ, Spirtas R, Gamble JF, Tousey PM (1976) Mortality among rubber workers: Relationship to specific jobs. J Occup Med 18:178–185.

Meinhardt T, Young R, Hartle R (1978) Epidemiologic investigations of styrene-butadiene rubber production and reinforced plastic production. Scand J Work Environ Hlth 8(4).250–259

Murray FJ, John JA, Haberstoh HD, et al. (1976) Teratologic evaluation of styrene monomers administered rats by gavage. Dow Chemical Study for Manufacturing Chemical Association. August 26.

Murray FJ, John JA, Balmer MF, Schwetz BA (1978) Teratologic evaluation of styrene given to rats and rabbits by inhalation or by gavage. Toxicology 11(4): 335–343.

NCI (1979) National Cancer Institute. Bioassay of styrene for possible carcinogenicity. National Cancer Institute carcinogenesis technical report series no. 185. NCI, Research Triangle Park, NC.

NIOSH (1983) National Institute for Occupational Safety and Health. Criteria for a recommended standard ... occupational exposure to styrene. Department of Health and Human Services (DHHS), NIOSH Publication no. 83-119. US DHHS, Washington, DC.

Odkvist LM, Larsby B, Tham R, Ahlfeldt H, Anderson B, Eriksson B, Liedgren SRC (1982) Vestibulo-oculomotor disturbances in humans exposed to styrene. Acta Oto-Laryngol 94(5-6): 487–493.

Ohtsuji M, Ikeda M (1971) Metabolism of styrene in the rat and the stimulatory effect of phenobarbital Toxicol Appl Pharmacol 18(2): 321–328.

Pellizzari ED, Hartwell TD, Harris BSH, Waddell RD, Whitaker DA, Erickson MD (1982) Purgeable organic compounds in mother's milk. Bull Environ Contam Toxicol 28(3): 322–328.

Plotnick HB, Weigel WW (1979) Tissue distribution and excretion of ^{14}C-styrene in male and female rats. Res Commun Chem Pathol Pharmacol 24(3): 515–524.

Ponomarkov VI, Tomatis L (1978) Effects of long-term oral administration of styrene to mice and rats. Scand J Work Environ Hlth 4 (Supplement 2): 127–135.

Ponomarkov V, Cabrai J, Wahrendorf J, Galendo D (1984) A carcinogenicity study of styrene-7,8-oxide in rats. Cancer Lett 24: 95–101.

Quast JF, Kalvins RP, Olson KJ, et al. (1978) Results of a toxicity study in dogs and teratogenicity studies in rabbits and rats administered monomeric styrene. Toxicol Appl Pharmacol 45: 293–294.

Quast JF, Humiston CG, Kalvins RV, Olson KJ, McCollister SB, Wade CE, Beyer JE, Schwetz BA (1979) Results of a toxicity study of monomeric styrene administered to beagle dogs by oral intubation for 19 months. Toxicology Research Laboratory, Health and Environmental Sciences, Dow Chemical Co., Midland, MI. Final report. Submitted to USEPA Office of Health and Environmental Assessment, Cincinnati, OH.

Ramsey JC, Young JD (1978) Pharmacokinetics of inhaled styrene in rats and humans. Scand J Work and Hlth 4(Supplement 2): 84–91.

Ramsey JC and Young JD (1980) Comparative pharmacokinetics of inhaled styrene in rats and humans. In: Proceedings 10th conference on environmental toxicology. November, 1979. AFAMRL-TR-79-121. Wright Patterson Air Force Base, OH. pp 103–117.

Ramsey JC, Young JD, Karbowski RJ, Chenoweth MB, McCarty LP, Braun WH (1980) Pharmacokinetics of inhaled styrene in human volunteers. Toxicol Appl Pharmacol 53(1): 54–63.

Sikov MR, Cannon WC, Carr DB, Miller RA, Montgomery LF, Phelps DW (1981) Teratologic assessment of butylene oxide, styrene oxide and methyl bromide. Study performed by Battelle Pacific Northwest Laboratory, Richland, WA, for National Institute of Occupational Safety and Health (NIOSH), Division of Biochemical and Behavioral Science, Experimental Toxicology Branch, Washington, DC. Department of Health and Human Services (DHHS) (NIOSH) Publication no. 81-124.

Smith AH, Ellis L (1977) Styrene butadiene rubber synthetic plants and leukemia (letter to the editor). J Occup Med 19(7):441.

Spencer HC, Irish DD, Adams EM, Rowe VK (1942) The response of laboratory animals to monomeric styrene. J Ind Toxicol 24(10):295–301.

Srivastava SP, Das M, Mushtaq M, Chandra SV, Seth PK (1982) Hepatic effects of orally administered styrene in rats. J Appl Toxicol 2(4):219–222.

Stewart RD, Dodd HC, Baretta ED, Schaffer AW (1968) Human exposure to styrene vapor. Arch Environ Hlth 16(5):656–662.

Teramoto K, Horiguchi S (1979) Absorption, distribution and elimination of styrene in man and experimental animals. Arch Hig Rada Toksikol 30(Supplement):431–439.

USEPA[a] (1988) Draft health effects criteria document for styrene. USEPA Office of Drinking Water, Washington, DC.

USEPA (1986) Guidelines for carcinogen risk assessment. Fed Regis 51(185):33992–34003. September 24.

USEPA[a] (1985a) Method 503.1. Volatile aromatic organic compounds in water by purge and trap gas chromatography. USEPA Environmental Monitoring and Support Laboratory (EMSL), Cincinnati, OH.

USEPA[a] (1985b) Method 524.1. Volatile organic compounds in water by purge and trap gas chromatography/mass spectrometry. USEPA Environmental Monitoring and Support Laboratory (EMSL), Cincinnati, OH.

USEPA[a] (1985c) Draft technologies and costs for removal of synthetic organic chemicals from potable water supplies. USEPA Science and Technology Branch, Criteria and Standards Division, Office of Drinking Water (ODW), Washington, DC.

Vainio H, Paakonen R, Ronnholm K, Raunio V, Pelkonen O (1976) A study on the mutagenic activity of styrene and styrene oxide. Scand J Work Environ 3:147–151.

Withey JR, Collins PG (1977) Pharmacokinetics and distribution of styrene monomer in rats after intravenous administration. J Toxicol Environ Hlth 3(5–6):1011–1120.

Withey JR, Collins PG (1979) The distribution and pharmacokinetics of styrene monomer in rats by the pulmonary route. J. Environ Pathol Toxicol 2(6):1329–1342.

Wolf MA, Rowe VK, McCollister DD, Hollingsworth RL, Oyen F (1956) Toxicological studies of certain alkylated benzenes and benzene. Arch Ind Hlth 14:387–398.

2,3,7,8-Tetrachlorodibenzo-*p*-Dioxin[a]

I. General Information

2,3,7,8-Tetrachlorodibenzo-*p*-dioxin, also known as dioxin; TCDBD; TCDD; 2,3,7,8-tetrachlorodibenzodioxin; 2,3,7,8-tetrachlorodibenzo-1,4-dioxin; and 2,3,7,8-TCDD, has no commercial uses (USEPA 1985a). Its structure and properties are shown in Table 1.

TCDD is a synthetic chemical that has no natural sources. It is not produced directly but is formed as a by-product in the manufacture of several chlorinated phenolic compounds. It can also be found in fly ash and flue gases of incinerators.

TCDD, which is extremely resistant to degradation once adsorbed onto soil, has a reported half-life of 10 to 12 yr. It has a low water solubility and binds readily to soil. It has been shown to migrate slowly in soil. Bioaccumulation of TCDD has been demonstrated in fish and mammals.

TCDD has not been included in drinking water surveys. Because of its limited solubility, it is not expected to occur at detectable levels in either ground or surface water. It may be present at low levels in some surface waters, where it is probably bound to suspended materials. TCDD was found in a number of freshwater fish at levels ranging from 1 to 695 ng/kg. It also was detected at low levels in rice treated with phenolic herbicides and in the fat of animals that grazed on pasture treated with phenolic herbicides. Because of the physical characteristics of TCDD, diet is expected to be a more significant route of exposure than drinking water; however, the available data are insufficient to evaluate the actual levels of exposure by either route (USEPA 1984a).

II. Pharmacokinetics

A. Absorption

Gavage treatment with single or repeated doses of TCDD in oil resulted in absorption of approximately 50% of an unspecified dose administered to guinea pigs (Nolan et al. 1979). Absorption in rats after administration of

[a] Based on USEPA (1985a), except for Section I which is based on USEPA (1984a)

Table 1. General information and properties of 2,3,7,8-tetrachlorodibenzo-*p*-dioxin

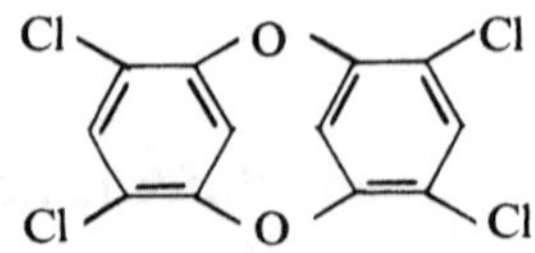

Property	Value
Chemical Abstracts Service (CAS) number	1746-01-6
Chemical formula	$C_{12}H_4Cl_4O_2$
Molecular weight	321.9
Physical state	Colorless solid, needle-shape
Boiling point	412.2°C[a]
Melting point	303 to 305°C
Density	1 827
Vapor pressure	3.5×10^{-9} mm Hg (30 1°C)[a]
Specific gravity	1.827
Water solubility	7.9×10^{-3} μg/L[b]
Log octanol water/partition coefficient	6.15
Odor threshold	—
Taste threshold	—

Source USEPA (1985a)

[a] Schroy et al (1985), converted from 4 68 $\times$ 10^{-7} pascals

[b] Adams and Blaine (1985)

1 or 50 μg/kg was 70 to 83% (Piper et al. 1973; Rose et al. 1976). Comparable oral absorption was reported in hamsters (650 μg/kg) (Olson et al. 1980a). Absorption of a single oral dose of [^{3}H]TCDD, 1.14 ng/kg, in corn oil by a male human volunteer was estimated to be 88.5% (Poiger and Schlatter 1986).

Dietary administration of 0.5 or 1.4 μg TCDD/kg/d for 42 d resulted in reduced gastrointestinal absorption in rats; approximately 50 to 60% of the administered dose was absorbed (Fries and Marrow 1975).

Diamond Shamrock (1985) noted that oral absorption of TCDD in animals given contaminated soil was greater when the soil contained oil than when oil was not present.

Percutaneous absorption of TCDD (26 ng) in rats is approximately 40% of the absorption of an equivalent dose administered orally (Poiger and Schlatter 1980).

Inhalation absorption of TCDD has not been studied (USEPA 1985a).

B. Distribution

Poiger and Schlatter (1986) detected concentrations of 3.0 and 2.8 ppt of [^{3}H]TCDD in adipose tissue at 10 and 69 d, respectively, after treatment.

Tissue distribution after oral or intraperitoneal (ip) administration of TCDD to rats appears to be preferential to liver and adipose tissue (Fries and Marrow 1975; Kociba et al. 1978b; Rose et al. 1976; Van Miller et al. 1976). Other tissues show substantially lower concentrations of TCDD. Soon after treatment, the liver may have TCDD concentrations of three (Kociba et al. 1978a) to five (Rose et al. 1976) times that of adipose tissue. It was suggested that male rats accumulate TCDD in the liver more efficiently than female rats (Fries and Marrow 1975). Tissue distribution in mice (Manara et al. 1982) and hamsters (Olson et al. 1980a) is similar to that in rats.

Monkeys, however, appear to accumulate TCDD in adipose tissue to a greater extent than in the liver (McNulty et al. 1982; Van Miller et al. 1976). Two yr after a single oral dose to a monkey, TCDD concentrations were 100 and 15 ppt in adipose tissue and liver, respectively (McNulty et al. 1982). Tissue distribution in guinea pigs is similar to that in monkeys (Gasiewicz and Neal 1979; Nolan et al. 1979) because levels in adipose tissue exceed those in the liver.

Evidence that TCDD accumulates in the adipose tissue of exposed humans was presented by Young et al. (1983), who reported levels of 3 to 99 ppt in the adipose tissue of veterans of the U.S. armed forces who claimed health problems related to Agent Orange.

Fetal distribution of TCDD has been studied in rats (Moore et al. 1976) and mice (Nau and Bass 1981; Nau et al. 1982). Levels of TCDD were low in rat fetuses on d 14 and 18 of gestation and appeared to be distributed evenly in all fetal tissues. On d 21 of gestation, the fetal liver showed a marked affinity for TCDD (Moore et al. 1976). TCDD was distributed to the fetuses of mice following oral, intraperitoneal, or subcutaneous administration (Nau et al. 1982). Maximum fetal concentrations occurred on d 9 and 10 of gestation; lower fetal concentrations were observed on d 11 through 18 of gestation, coincident with placentation. The fetal liver had less affinity for the compound than did the maternal liver.

Ryan et al. (1985) reported TCDD levels of 5 to 10 ppt in Canadian autopsy samples of human adipose tissue. Higher levels of other dioxins were also found.

C. Metabolism

In an early metabolism study, Vinopal and Casida (1973) reported in vivo and in vitro data indicating that polar metabolites of TCDD were not produced by the mouse species they studied. In rats, however, hydroxylation and conjugation with glucuronide and sulfate have been demonstrated (Olson et al. 1983; Poiger and Schlatter 1979; Poiger et al. 1982). Glucuronide conjugates tend to predominate in bile (Poiger and Schlatter 1979) and sulfate conjugates are found in urine (Olson et al. 1983).

Poiger and Schlatter (1979) stated that metabolism of TCDD proceeds slowly in the liver. Neal et al. (1982) demonstrated that the rate of hepatic metabolism was enhanced by activated cytochrome P-450 mono-oxygenase. It was suggested that metabolism of TCDD proceeds by the formation of reactive epoxide intermediates (Poland and Glover 1979). Dechlorination was demonstrated by Olson et al. (1983) and Sawahata et al. (1982), who identified trichlorodibenzo-*p*-dioxin and dichlorodibenzo-*p*-dioxin as metabolites in in vitro rat hepatocyte systems. Six major metabolites were identified in the bile of dogs (Poiger et al. 1982); hydroxylated conjugates of TCDD, trichlorodibenzo-*p*-dioxin, and dichlorodibenzo-*p*-dioxin predominated.

Although these metabolite profiles are consistent with the proposed arene oxide intermediate, the covalent interaction of TCDD with cellular macromolecules is minimal (USEPA 1985a,b).

D. Excretion

TCDD excretion exhibits first-order kinetics, especially in rats. Excretion in guinea pigs, however, may be a zero-order process (Gasiewicz and Neal 1979). Measures of the half-life for body elimination vary considerably. In hamsters, the species least sensitive to the toxic effects of TCDD, elimination half-lives ranged from 10 to 15 d (Olson et al. 1980a). In mice, the half-life is 11 to 24 d (Gasiewicz et al. 1983a,b). In rats, the half-life is 17 to 31 d (Allen et al. 1975; Piper et al. 1973; Rose et al. 1976). In guinea pigs, the half-life is 22 to 30 d (Gasiewicz and Neal 1979; Nolan et al. 1979). In one strain of mice (DBA/2J), the half-life for elimination was approximately 24 d, about twice as long as in other strains tested by Gasiewicz et al. (1983a,b). These authors noted that DBA/2J mice had a greater tendency to accumulate TCDD in adipose tissue, resulting in slower body elimination. Half-lives for body elimination of TCDD have not been calculated for monkeys, but the tendency of this species to accumulate TCDD in adipose tissue may also result in slow body elimination (Van Miller et al. 1976).

Recently, Olson and Bittner (1983) examined the elimination of TCDD in rats over a longer period than in the studies previously summarized. These investigators found TCDD elimination to be biphasic with half-lives of approximately 7 d for the initial rapid phase and approximately 75 d for the slower phase. The half-life in the slower phase was probably related to release of TCDD from body fat. McNulty et al. (1982) estimated the half-life for elimination from fat in monkeys to be approximately 1 yr.

Poiger and Schlatter (1986) found that 11.5% of [^{3}H]TCDD was excreted in feces during the first 3 d after treatment; no ^{3}H activity was found in urine. These investigators estimated an elimination half-life of 4.95 yr for [^{3}H]TCDD in humans.

Feces seem to be the major route of elimination of TCDD-derived radioactivity in rats (Piper et al. 1973; Allen et al. 1975; Rose et al. 1976; Van Miller et al. 1976), guinea pigs (Gasiewicz and Neal 1979), and mice (Gasiewicz et al. 1983a,b). Urinary excretion, a minor pathway in these species, accounted for < 1 to 28% of total excreted radioactivity; fecal excretion accounted for 72 to > 99% of eliminated radioactivity. In hamsters, urinary excretion (41%) was more significant than in other species relative to fecal excretion (59%) (Olson et al. 1980a). This was also true in DBA/2J mice that had preferential accumulation of TCDD in body fat (Gasiewicz et al. 1983a,b).

Failure to detect metabolites of TCDD in liver and fat (Olson et al. 1983) indicates that elimination of metabolites occurs rapidly and the rate of elimination is governed primarily by the rate of hepatic metabolism.

III. Health Effects

A. Humans

Acute or chronic exposure to TCDD (usually in combination with other substances) may result in chloracne, altered liver function, hematologic lesions, porphyria cutanea tarda, hyperpigmentation, hirsuitism, and neural degeneration in the extremities (USEPA 1985a). Stevens (1981) estimated the minimum cumulative toxic dose of TCDD in humans to be 0.1 μg/kg.

Rowe (1968) found a dose–response relationship for TCDD and chloracne in humans acutely exposed to topical applications. The toxic effects of chloracne from exposure to TCDD may persist for many years, although other effects are apparently reversible sooner.

Epidemiologic studies have failed to demonstrate a convincing connection between TCDD exposure and spontaneous abortions or fetal malformations in humans. Some evidence of cytogenetic damage was reported in humans exposed to chemicals contaminated with TCDD, but negative results were also reported. Exposures in these studies were not quantitated; therefore, other chemicals cannot be ruled out as causative agents (USEPA 1985a).

Swedish case-control studies provide limited evidence for the carcinogenicity of phenoxy acids or chlorophenols or both in humans. Evidence of the carcinogenicity of TCDD present as an impurity within these chemicals is only suggestive, however, in the presence of confounding effects of phenoxy acids and chlorophenol (USEPA 1985a).

B. Animals

1. Short-Term Exposure. Sensitivity to the acute toxicity of TCDD varies widely among species. LD_{50}s range from 0.6 μg/kg for male guinea pigs to > 5,000 μg/kg for male hamsters (Henck et al. 1981; Olson et al. 1980b; Schwetz et al. 1973). The toxic manifestations seem to be the same whether the TCDD is given as a single oral dose or as a limited number of multiple treatments; death occurs 5 to 45 d after treatment. Lethal exposures result in weight loss and thymic atrophy. In some species, particularly rats and mice, extensive liver damage has been observed (Gupta et al. 1973). Usually, no specific cause of death can be identified, although extensive hemorrhage was implicated in mice (Vos et al. 1974).

In rats, single high doses (200 μg/kg) produced liver necrosis (Jones and Butler 1974), and lower doses (5 and 25 μg/kg) resulted in fatty changes in the liver and proliferation of endoplasmic reticulum (Fowler et al. 1973). Other effects seen in some species included induction of microsomal enzymes, degeneration of plasma membranes with loss of adenosine triphosphatase (ATPase) activity, a decreased ability to excrete some xenobiotics in bile, porphyria, altered gastrointestinal absorption of some nutrients, and decreased blood cellularity (USEPA 1985a). Turner and Collins (1983) found treatment-related liver lesions in guinea pigs given single gavage doses of TCDD at 0.1 μg/kg and higher.

TCDD is an immunotoxin in laboratory animals, predominantly affecting cell-mediated immunity. Hypersensitivity, adverse effects on the thymus, and increased sensitivity to antigens demonstrate the immunotoxic potential of TCDD. Weanling rodents show greater susceptibility to immune effects than do adults (USEPA 1985a).

2. Long-Term Exposure. In rats and mice, the liver appears to be the most sensitive organ after chronic or subchronic exposure. Hepatotoxicity develops following a long induction period, and changes persist for lengthy periods after termination of exposure (Goldstein et al. 1982; King and Roesler 1974).

Hepatic lesions and other toxic signs were observed in the following studies in rats and mice. In subchronic studies, No-Observed-Adverse-Effect Levels (NOAELs) of 0.01 μg/kg/d (Kociba et al. 1976) and 0.5 μg/kg/wk (NTP 1980) were reported for rats. A NOAEL of 2 μg/kg/wk was identified for female mice and a Lowest-Observed-Adverse-Effect Level (LOAEL) of 1 μg/kg/wk was reported for male mice in a subchronic study (NTP 1980). A NOAEL of 0.001 μg/kg/d, a LOAEL of 0.01 μg/kg/d, and an effect level of 0.1 μg/kg/d were reported for rats after chronic dietary exposure (Kociba et al. 1978a,b, 1979; NTP 1980). Toth et al. (1978, 1979) observed toxic effects in mice at doses as low as 0.007 μg/kg/wk given for 1 yr by gavage. Gavage doses for 2 yr led to toxic hepatitis (NOAEL: 0.05 μg/kg/wk, LOAEL: 0.5 μg/kg/wk in rats; LOAEL: 0.5 μg/kg/d in male mice; and LOAEL: 2.0 μg/kg/wk in female mice) (NTP 1980).

DeCaprio et al. (1986) fed TCDD in the diet for 90 d to male and female Hartley guinea pigs and found NOAELs of 0.12 and 0.61 mg/kg/d in males and 0.68 mg/kg/d in females. They also observed decreased body weight gain, increased relative liver weights, decreased relative thymus weights, and hepatocellular cytoplasmic inclusion bodies at 4.90 mg/kg/d (males) and 4.86 mg/kg/d (females). Mortality and other effects were observed at 26 mg/kg/d (males) and 31 mg/kg/d (females).

3. Reproductive Effects. Adverse effects of TCDD on reproduction in rats exposed through diet were observed by Murray et al. (1979) and are described under Lifetime Health Advisory (HA) below.

4. Developmental Effects. TCDD has been demonstrated to be teratogenic in mice. The most common malformations are cleft palate and kidney anomalies; however, other malformations have been observed. With an effect level of 1 μg/kg/d, TCDD is the most potent teratogen known. At higher doses, TCDD has a marked fetotoxic effect, as measured by decreased fetal weight and increased fetal toxicity. Hemorrhagic gastrointestinal tract has been associated with TCDD fetal toxicity (USEPA 1985a).

Poland and Glover (1980) have demonstrated that cleft palate in mice after TCDD treatment is related to the presence of Ah receptor.

In rats, TCDD is consistently observed to produce fetotoxic responses. In this species, the most common fetal anomalies observed are edema, hemorrhage, and malformation of the kidneys, with effects observed at doses of > 0.01 μg/kg/d. In addition some evidence exists that TCDD can induce microsomal enzymes in the fetus exposed in utero. Enzyme induction may be accompanied by damage to the fine structure of liver cells; however, other reports indicate that enzyme induction occurs only after birth and exposure to TCDD through maternal milk. As in mice, hemorrhagic gastrointestinal tracts were observed in rat fetuses exposed in utero to TCDD (USEPA 1985a).

Rabbits and monkeys are also susceptible to the fetotoxic effects of TCDD; however, studies of these species have been too limited to clearly demonstrate a teratogenic response or to define a threshold dose for fetotoxicity (USEPA 1985a).

5. Mutagenicity. In vivo and in vitro mutagenicity tests produced inconclusive evidence of the mutagenicity of TCDD (USEPA 1985a).

Early reports indicated that TCDD was mutagenic in *Salmonella typhimurium* strain TA1532 (Hussain et al. 1972; Seiler 1973). However, later attempts to confirm these results were unsuccessful (Nebert et al. 1976; McCann 1978; Gilbert et al. 1980; Geiger and Neal 1981). TCDD was reported to be mutagenic to *Escherichia coli* in vitro (Hussain et al. 1972), to *Saccharomyces cerevisiae* in vitro, and in a host-mediated assay (Bronzetti et al. 1983). Covalent interactions with nucleic acids are minimal if they occur at all (Kondorosi et al. 1973; Poland and Glover 1979). Only marginal effects were observed on the incidence of chromosomal aberrations in vivo (Green and Moreland 1975). Results of a test for unscheduled deoxyribonucleic acid

(DNA) synthesis in cultured male rat hepatocytes were negative (Althaus et al. 1982). Loprieno et al. (1982) reported TCDD as clastogenic in mice in vivo, negative for cytogenetic effects in vivo, and negative for unscheduled DNA synthesis in a human cell line in vitro. Hay (1983) reported TCDD as mutagenic in a baby hamster kidney cell transformation assay.

6. Carcinogenicity. Several bioassays have demonstrated this compound to be a potent carcinogen in rats and mice. Thyroid adenomas and carcinomas, hepatocellular carcinomas, carcinomas of the tongue and hard palate, and adrenal adenomas have been induced by TCDD in rats and mice (Kociba et al. 1978a; NTP 1980; Toth et al. 1979).

Significant ($p < 0.05$) neoplastic effects were evident at dietary levels of 0.01 and 0.1 μg/kg/d but not at 0.001 μg/kg/d in a 2-yr study with Sprague–Dawley rats by Kociba et al. (1978a). Osborne–Mendel rats were given TCDD in a corn oil and acetone mixture twice weekly for total doses of 0.01, 0.05, and 0.5 μg/kg/wk for 2 yr. Significant ($p < 0.05$) increases in thyroid tumors in mid- and high-dose males and liver tumors in high-dose males were observed (NTP 1980). In a study of B6C3F1 mice, significant ($p < 0.05$) increases in tumors occurred in the liver of high-dose males and females and in the thyroid of high-dose males (NTP 1980). The dosages for males were identical to those of Kociba et al. (1978a). Females received total weekly doses of 0.04, 0.2, and 2.0 μg/kg.

IV. Quantification of Toxicological Effects[a]

A. One-Day Health Advisory

Turner and Collins (1983) administered single oral doses of TCDD at 0.1, 0.5, 2.5, 12.5, or 20 μg/kg in aqueous methylcellulose to groups of four to seven female guinea pigs. Survivors were sacrificed at 42 d after dosing and examined for histopathologic changes in the liver. Four of the seven animals in the highest dose group and one of five in the 12.5-μg/kg group died before the end of the observation period. Mild histopathologic changes, including steatosis (fatty change), focal necrosis, and cytoplasmic degeneration, were noted in animals from all treated groups but not in controls. The authors indicated that quantitative differences among the dosage groups were not detectable by light microscopy.

A LOAEL of 0.1 μg/kg can be derived from the study of Turner and Collins (1983) for calculating a One-Day HA, using an uncertainty factor (UF) of 1,000 for an animal LOAEL. This UF consists of two 10-fold factors to account for both intraspecies and interspecies variability for the toxicity of

[a] See Introduction for a discussion of methodology

TCDD in the absence of chemical-specific data and an additional 10-fold factor because the HA is based on a LOAEL and not a NOAEL.

For a 10-kg child, the One-Day HA is calculated as follows:

$$\text{One-Day HA} = \frac{(0.1\ \mu g/kg/d)\ (10\ kg)}{(1{,}000)\ (1\ L/d)} = 0.001\ \mu g/L$$

B. Ten-Day Health Advisory

Because of the demonstrated sensitivity of the guinea pig to acute toxicity of TCDD, the Ten-Day HA is derived by dividing the One-Day HA by 10. The Ten-Day HA is, therefore, 0.0001 μg/L.

C. Longer-Term Health Advisory

The three-generation reproduction study in rats by Murray et al. (1979) (described under Lifetime HA) was selected to calculate the Longer-Term HA because the study was of appropriate duration, the route of exposure was diet, and the study examined reproductive effects. Comparison with the other studies in which different treatment protocols were used suggests that 0.001 μg/kg/d, a LOAEL for adverse effects on pups and dams (Murray et al. 1979), would be protective against the toxic effects seen in the other studies. Although DeCaprio et al. (1986) found NOAELs of 0.61 and 0.68 ng/kg/d in a 90-d guinea pig study, this dose is slightly less than the LOAEL of 0.001 μg/kg/d (1 ng/kg/d) in another species, which in turn is less than the LOAEL of 4.86 ng/kg/d in the DeCaprio et al. (1986) study.

Using an uncertainty factor of 1,000 for an animal LOAEL and a LOAEL of 0.002 μg/kg/d based on adverse effects on reproduction in rats, a Longer-Term HA can be calculated as follows:

For a 10-kg child:

$$\text{Longer-Term HA} = \frac{(0.001\ \mu g/kg/d)\ (10\ kg)}{(1{,}000)\ (1\ L/d)} = 0.00001\ \mu g/L$$

For a 70-kg adult:

$$\text{Longer-Term HA} = \frac{(0.001\ \mu g/kg/d)\ (70\ kg)}{(1{,}000)\ (2\ L/d)} = 0.000035\ \mu g/L$$

D. Lifetime Health Advisory

For comparison with cancer-based criteria, the USEPA developed a presumed safe daily intake level based on noncarcinogenic effects (USEPA 1984b). For consistency, the rationale used by USEPA (1984b) for the calculation of this value is also used here for calculation of the Drinking Water Equivalent Level (DWEL). The rationale is as follows:

TCDD displays an unusually high degree of reproductive toxicity. It is teratogenic, fetotoxic, and reduces fertility. In a three-generation reproductive study, Murray et al. (1979) reported a reduction in fertility after daily dosing at 0.1 or 0.01 μg/kg in the F_1 and F_2 generations of Sprague–Dawley rats. Although Murray et al. (1979) considered the lowest dose tested, 0.001 μg/kg, to be a NOEL, a reevaluation of these data by Nisbet and Paxton (1982) using different statistical methods detected a reduction in gestation index, decreased fetal weight, increased liver-to-body weight ratio, and increased incidence of dilated renal pelvis at the 0.001-μg/kg dose. The reevaluated data suggest that equivocal adverse effects were seen at 0.001 μg/kg/d and that this dose is a LOAEL. Schantz et al. (1979) found reductions in fertility and various other toxic effects in rhesus monkeys fed a 50-ppt TCDD diet for 20 mon. This corresponds to a calculated dose of 0.0015 μg/kg/d. These results suggest that monkeys may be somewhat more sensitive than rats, since the effects in monkeys were more severe and not equivocal. Since the data from the limited study of Schantz et al. (1979) support the findings by Murray et al. (1979) it seems reasonable to use 0.001 μg/kg/d as a LOAEL.

Using this LOAEL, the DWEL is calculated as follows:

Step 1: Determination of the Reference Dose (RfD)

$$\text{RfD} = \frac{(0.001\ \mu\text{g/kg/d})}{(1{,}000)} = 1 \times 10^{-6}\ \mu\text{g/kg/d}$$

Step 2: Determination of the Drinking Water Equivalent Level (DWEL)

$$\text{DWEL} = \frac{(1 \times 10^{-6}\ \mu\text{g/kg/d})\ (70\ \text{kg})}{(2\ \text{L/d})} = 0.000035\ \mu\text{g/L}$$

Step 3: Determination of Lifetime Health Advisory

A Lifetime HA is *not* recommended because TCDD is classified in Group B2: probable human carcinogen (see *Evaluation of Carcinogenic Potential* below). For comparison purposes, the cancer risk associated with the DWEL was determined.

The estimated excess cancer risk associated with lifetime exposure to drinking water containing TCDD at 3.5×10^{-5} μg/L is approximately 2×10^{-4}. This estimate represents the upper 95% confidence limit from extrapolations prepared by the USEPA Carcinogen Assessment Group (CAG), using the linearized, multistage model. The actual risk is unlikely to exceed this value, but considerable uncertainty exists as to the accuracy of risks calculated by this method.

E. Evaluation of Carcinogenic Potential

Applying the criteria described in USEPA guidelines for assessment of carcinogenic risk (USEPA 1986), TCDD may be classified in Group B2:

probable human carcinogen. This category is for agents for which there is inadequate evidence of carcinogenicity from human studies and sufficient evidence from animal studies. The International Agency for Research on Cancer classified TCDD in category 2B: sufficient animal evidence and inadequate human evidence of carcinogenicity (IARC 1982).

Cancer potency estimates were derived using the multistage model and the tumor data on female rats in the chronic feeding study by Kociba et al. (1978a) (USEPA 1985a,b).

The 95% upper-limit carcinogenic potency factor for humans, $q_1{}^*$, is 1.56×10^5 (mg/kg/d)$^{-1}$. For a 70-kg human drinking 2 L/d of water, the water concentration should be 2.2×10^{-6} μg/L to maintain the upper-limit individual lifetime cancer risk at 10^{-5}. Water concentrations corresponding to excess cancer risks of 10^{-4} and 10^{-6} are 2.2×10^{-5} and 2.2×10^{-7} μg/L, respectively.

Maximum likelihood estimates (MLEs) and 95% upper limits (ULs) of cancer risks have been calculated using the multistage model (USEPA 1985b). For example, at 1×10^{-3} ng/kg/d or 0.035 ng/L, cancer risk estimates are 1.1×10^{-4} (MLE) and 1.5×10^{-4} (UL); at 1×10^{-2} ng/kg/d, cancer risk estimates are 1.1×10^{-3} (MLE) and 1.5×10^{-3} (UL).

The USEPA CAG estimated cancer risks with other models in addition to the multistage model (USEPA 1985b). For example, 1×10^{-3} ng/kg/d lifetime exposure was associated with additional risks (95% upper confidence limit) of 1.5×10^{-4} by the multistage and one-hit models, 2.9×10^{-3} by the Weibull model, and 7.5×10^{-8} by the log-probit model, using the Kociba (1978a) analysis of the data. In the interest of consistency of approach and to provide an upper limit on potential cancer risk, the USEPA has recommended use of the linearized, multistage approach.

V. Other Criteria, Guidance, and Standards

The USEPA has established criteria of 1.3×10^{-7}, 1.3×10^{-8}, or 1.3×10^{-9} μg/L for TCDD in ambient waters, based on an assumed daily consumption of 6.5 g of contaminated fish and shellfish and 2 L drinking water (USEPA 1984b). Under these conditions, 94.2% of the total exposure to TCDD would result from the consumption of aquatic organisms. The recommended levels correspond to estimated human lifetime excess cancer risks of 10^{-5}, 10^{-6}, or 10^{-7}, respectively. These values are considerably lower than the HAs for drinking water, thus reflecting the high bioaccumulation potential of this compound in aquatic species.

The U.S. Food and Drug Administration advised that fish containing > 50 ppt of TCDD should not be consumed and fish containing > 25 ppt but < 50 ppt should not be consumed more than twice a month (USFDA 1983). A limit of 20 ppt in Lake Ontario commercial fish imported by Canada to the U.S. (NRCC 1981) reflects USFDA's recommendation.

An Acceptable Daily Intake (ADI) of 10^{-4} μg/kg/d for TCDD was proposed previously by the NAS Safe Drinking Water Committee (NAS 1977). This ADI was based on a 13-wk feeding study in rats by Kociba et al. (1976) and was proposed before convincing evidence for the carcinogenicity of TCDD had accumulated.

VI. Analytical Methods

Dioxin levels are determined by a gas chromatographic (GC) and mass spectrometer (MS) technique (Method 613) (USEPA 1984c). In this method, a 1-L sample is spiked with an internal standard of labeled dioxin and extracted with methylene chloride using a separatory funnel. The methylene chloride extract is exchanged to hexane during concentration to a volume of approximately 1 mL. The extract is analyzed by capillary column GC-MS to separate and measure dioxin. The detection limit for this method is dependent on the nature of interferences, but it is estimated to be about 0.02 μg/L.

VII. Treatment Technologies

Because of its high toxicity and low potential for occurrence in drinking water, little information is available on the removal of dioxin from drinking water. The small amount of empirical evidence currently available indicates that granular-activated carbon (GAC) is likely to be the most reasonable treatment.

While looking for a method to concentrate polychlorinated dibenzo-*p*-dioxins and dibenzofurans, scientists from the U.S. Fish and Wildlife Service fish-pesticide research laboratory in Columbia, Missouri, found that TCDD is extremely difficult to recover from GAC once it has been adsorbed. Subsequent pilot-scale tests of carbon adsorption of Agent Orange, a 50-50 mixture of the acid esters of 2,4,5-T (2,4,5-trichlorophenoxy acetic acid) and 2,4-D (2,4-dichlorophenoxy acetic acid), reduced an initial concentration of 10 mg/L dioxin in the herbicide to a final concentration of < 0.1 mg/L (Chemical Engineering and News 1977). Details of the adsorption test were not reported by the authors. Based on these data and the reported low water solubility of 0.2 μg/L dioxin in water (Bollen and Norris 1979), GAC adsorption of dioxin from water appears potentially feasible.

References

Adams WJ, Blaine KM (1985) A water solubility of 2,3,7,8-TCDD. Monsanto Company, St. Louis, MO. Dioxin '85—5th international symposium on chlorinated dioxins and related compounds, held in Bayreuth, West Germany, September 16–19, 1985.

Allen JR, Van Miller JP, Norback DH (1975) Tissue distribution, excretion, and biological effects of (14-C) tetrachlorodibenzo-*p*-dioxin in rats. Food Cosmet Toxicol 13(5): 501–505.

Althaus FR, Lawrence SD, Sattler GL, Longfellow DG, Pitot HC (1982) Chemical quantification of unscheduled DNA synthesis in cultured hepatocytes as an assay for the rapid screening of potential chemical carcinogens. Cancer Res 42: 3010–3015.

Bollen WB, Norris LA (1979) Influence of 2,3,7,8-tetrachlorodibenzo-*p*-dioxin on respiration in a forest floor and soil. Bull Environ Contam Toxicol 22: 648–652:

Bronzetti G, Zeiger E, Lee I, Suzuki K, Malling HV (1983) Mutagenicity study of TCDD and ashes from urban incinerator *in vitro* and *in vivo* using yeast D7 strain. Chemosphere 12: 549–553.

Chemical Engineering and News (1977) Method rids Agent Orange of TCDD contamination. CEN 55(11): 25.

DeCaprio AP, McMartin DN, O'Keefe PW, Rej R, Silkworth JB, Kaminsky LS (1986) Subchronic oral toxicity of 2,3,7,8-tetrachlorodibenzo-*p*-dioxin in the guinea pig. Comparisons with a PCB-containing transformer fluid pyrolysate. Fund Appl Toxicol 6: 454–463.

Diamond Shamrock[a] (1985) Letter to USEPA Office of Drinking Water (ODW), with comments on 2,3,7,8-TCDD Health Advisory document USEPA Office of Drinking Water, Washington, DC.

Fowler BA, Lucier GW, Brown HW, McDaniel OS (1973) Ultrastructural changes in rat liver cells following a single oral dose of TCDD Environ Hlth Perspect 5: 141–148.

Fries GF, Marrow GS (1975) Retention and excretion of 2,3,7,8-tetrachlorodibenzo-*p*-dioxin by rats. J Agric Food Chem 23(2): 265–269.

Gasiewicz TA, Neal RA (1979) 2,3,7,8-Tetrachlorodibenzo-*p*-dioxin tissue distribution, excretion, and effects on clinical chemical parameters in guinea pigs. Toxicol Appl Pharmacol 51(2): 329–340

Gasiewicz TA, Olson JR, Geiger LE, Neal RA (1983a) Absorption, distribution and metabolism of 2,3,7,8-tetrachlorodibenzo-*p*-dioxin (TCDD) in experimental animals. In: Tucker RE, Young AL, Gray AP (eds) Human and environmental risks of chlorinated dioxins and related compounds. Plenum Press, NY, pp 495–525.

Gasiewicz TA, Geiger LE, Rucci G, Neal RA (1983b) Distribution, excretion and metabolism of 2,3,7,8-tetrachlorodibenzo-*p*-dioxin in C57Bl/6J, DBA/2J, and B6D2F_1/J mice. Drug Metab Dispos 11(5): 397–403.

Geiger LE, Neal RA (1981) Mutagenicity testing of 2,3,7,8-tetrachlorodibenzo-*p*-dioxin in histidine auxotrophs of *Salmonella typhimurium*. Toxicol Appl Pharmacol 59(1) · 125–129.

Gilbert P, Saint-Ruf G, Poncelet F, Mercier M (1980) Genetic effects of chlorinated

[a] USEPA documents can be obtained by writing to the appropriate office within the U S Environmental Protection Agency, 401 M Street, S W , Washington, DC 20460 They may also be available by contacting the National Technical Information Service, 5285 Port Royal Road, Springfield, VA 22161 (Telephone 1-800-336-4700)

anilines and azobenzenes on *Salmonella typhimurium*. Arch Environ Contam Toxicol 9(5):533–541.

Goldstein JA, Linko P, Bergman H (1982) Induction of porphyria in the rat by chronic versus acute exposure to 2,3,7,8-tetrachlorodibenzo-*p*-dioxin. Biochem Pharmacol 31(8):1607–1613.

Green S, Moreland FS (1975) Cytogenetic evaluation of several dioxins in the rat. Toxicol Appl Pharmacol 33:161.

Gupta BN, Vos JG, Moore JA, Zinkl JG, Bullock BC (1973) Pathologic effects of 2,3,7,8-tetrachlorodibenzo-*p*-dioxin in laboratory animals. Environ Hlth Perspect 5:125–140.

Hay A (1983) The mutagenic properties of 2,3,7,8-tetrachlorodibenzo-*p*-dioxin. Abstracts of American Chemical Society National Meeting 23(2):14.

Henck JW, New MA, Kociba RJ, Rao KS (1981) 2,3,7,8-Tetrachlorodibenzo-*p*-dioxin: acute oral toxicity in hamsters. Toxicol Appl Pharmacol 59:405–407.

Hussain S, Ehrenberg L, Lofroth G, Gejvall T (1972) Mutagenic effects of TCDD on bacterial systems. Ambio 1:32–33.

IARC (1982) International Agency for Research on Cancer IARC monographs on the evaluation of the carcinogenic risk of chemicals to humans. Chemical industrial processes and industries associated with cancer in humans. Supplement 4, pp 238–243. IARC, Lyon, France.

Jones G, Butler WH (1974) A morphological study of the liver lesion induced by 2,3,7,8-tetrachlorodibenzo-*p*-dioxin in rats. J Pathol 112:93–97.

King ME, Roesler AR (1974) Subacute intubation study on rats with the compound 2,3,7,8-tetrachlorodioxin. NTIS[a] PB 257 677, p 27. National Technical Information Service, Springfield, VA.

Kociba RJ, Keeler PA, Park CN, Gehring PJ (1976) 2,3,7,8-tetrachlorodibenzo-*p*-dioxin results of a 13-week oral toxicity study in rats. Toxicol Appl Pharmacol 35:553–574.

Kociba RJ, Keyes DG, Beyer JE, Carreon RM, Wade CE, Dittenber DA, Kalnins RP, Frauson LE, Park CN, Barnad SD, Hummel RA, Humiston CG (1978a) Results of a two-year chronic toxicity and oncogenicity study of 2,3,7,8-tetrachlorodibenzo-*p*-dioxin in rats. Toxicol Appl Pharmacol 46(2):279–303.

Kociba RJ, Keyes DG, Beyer JE, Carreon RM (1978b) Toxicologic studies of 2,3,7,8-tetrachlorodibenzo-*p*-dioxin (TCDD) in rats. Toxicol Occup Med 4:281–287

Kociba RJ, Keyes DG, Beyer JE, Carreon RM, Gehring PJ (1979) Long-term toxicologic studies of 2,3,7,8-tetrachlorodibenzo-*p*-dioxin (TCDD) in laboratory animals. Ann NY Acad Sci 320:397–404.

Kondorosi A, Fedorcsak I, Solymosy F, Ehrenberg L, Osterman-Golkar S (1973) Inactivation of QBRNA by electrophiles. Mutat Res 17:149–161.

Loprieno N, Skrana I, Rusciano D, Lascialfari D, Lari T (1982) *In vivo* cytogenetic studies on mice and rats exposed to tetrachlorodibenzo-*p*-dioxin (TCDD). Book 5. Chlorinated dioxins and related compounds, impact on the environment, pp 419–428.

Manara L, Coccia P, Croci T (1982) Persistent tissue levels of TCDD in the mouse and their reduction as related to prevention of toxicity Drug Metab Rev 13(3):423–446.

McCann J (1978) Unpublished study (cited in Wassom et al. 1978)

McNulty WP, Nielsen-Smith KA, Lay JO Jr, Lippstrew DL, Kangas NL, Lyon PA, Gross ML (1982) Persistence of TCDD in monkey adipose tissue. Food Cosmet Toxicol 20:985–987.

Moore JA, Harris MW, Albro PW (1976) Tissue distribution of (^{14}C) tetrachlorodibenzo-*p*-dioxin in pregnant and neonatal rats. Toxicol Appl Pharmacol 37(1): 146–147.

Murray FJ, Smith FA, Nitschke KD, Humiston CG, Kociba RJ, Schwetz BA (1979) Three-generation reproduction study of rats given 2,3,7,8-tetrachlorodibenzo-*p*-dioxin (TCDD) in the diet. Toxicol Appl Pharmacol 50:241–251.

NAS (1977) National Academy of Sciences. Drinking water and health: part II. NAS, Washington, DC, pp 500–513.

Nau H, Bass R (1981) Transfer of 2,3,7,8-tetrachlorodibenzo-*p*-dioxin (TCDD) to the mouse embryo and fetus. Toxicology 20(4):299–308.

Nau H, Bass R, Neubert D (1982) Transfer of 2,3,7,8-tetrachlorodibenzo-*p*-dioxin (TCDD) to the mouse embryo, fetus, and neonate. In: Huntzinger O, Frei RW, Merian E, Pocchiari F (eds) Chlorinated dioxins and related compounds. Impact on the environment. Pergamon Press, NY, pp 325–337.

Neal RA, Olson JR, Gasiewicz TA, Geiger LE (1982) The toxicokinetics of 2,3,7,8-tetrachlorodibenzo-*p*-dioxin in mammalian systems. Drug Metab Rev 13:355–385.

Nebert D, Thorgiersson S, Felton J (1976) Genetic differences in mutagenesis, carcinogenesis and drug toxicity. In: de Serros F, Folets J, Bend J, Philpot R (eds) *In vitro* metabolic activation in mutagenesis testing. Elsevier/North Holland Biomedical Press, Amsterdam, pp 105–124.

Nisbet ICT, Paxton MB (1982) Statistical aspects of three-generation studies of the reproductive toxicity of TCDD and 2,4,5-T. Am Stat Vol 36(3):290–298.

Nolan RJ, Smith FA, Hefner JG (1979) Elimination and tissue distribution of 2,3,7,8-tetrachlorodibenzo-*p*-dioxin (TCDD) in female guinea pigs following a single oral dose. Toxicol Appl Pharmacol 48(1):A162.

NRCC (1981) National Research Council of Canada. Polychlorinated dibenzo-*p*-dioxins: criteria for their effects on man and his environment. NRCC/CNRC Associate Committee on Scientific Criteria for Environmental Quality, Ottawa, Canada. Publication no. NRCC 18574, ISSN 0316-0114. 251 pp.

NTP (1980) National Toxicology Program. Bioassay of 2,3,7,8-tetrachlorodibenzo-*p*-dioxin for possible carcinogenicity (gavage study). Carcinogenesis Testing Program, National Cancer Institute (NCI), National Institutes of Health (NIH), Bethesda, MD; NTP, Research Triangle Park, NC. U.S. Department of Health and Human Services Publication no. 82-1765.

Olson JR, Bittner WE (1983) Comparative metabolism and elimination of 2,3,7,8-tetrachlorodibenzo-*p*-dioxin (TCDD). Toxicologist 3:103.

Olson JR, Gasiewicz TA, Neal RA (1980a) Tissue distribution, excretion, and metabolism of 2,3,7,8-tetrachlorodibenzo-*p*-dioxin (TCDD) in the golden syrian hamster. Toxicol Appl Pharmacol 56:78–85.

Olson JR, Holscher MA, Neal RA (1980b) Toxicity of 2,3,7,8-tetrachlorodibenzo-*p*-dioxin in the golden syrian hamster. Toxicol Appl Pharmacol 55:67–78.

Olson JR, Gasiewicz TA, Geiger LE, Neal RA (1983) The metabolism of 2,3,7,8-tetrachlorodibenzo-*p*-dioxin in mammalian systems. In: Coulston R, Pocchiari F (eds) Accidental exposure to dioxins: human health aspects. Academic Press, NY, pp 81–100.

Piper WN, Rose RQ, Gebring PJ (1973) Excretion and tissue distribution of 2,3,7,8-tetrachlorodibenzo-*p*-dioxin in the rat. Environ Hlth Perspect 5:241–244.

Poiger H, Schlatter C (1979) Biological degradation of TCDD in rats. Nature 281(5733):706–707.

Poiger H, Schlatter C (1980) Influence of solvents and adsorbents on dermal and intestinal absorption of TCDD. Food Cosmet Toxicol 18(5):477–481.

Poiger H, Schlatter C (1986) Pharmacokinetics of 2,3,7,8-TCDD in man. Chemosphere (in press). Presented at the fifth international symposium on chlorinated dioxins and related compounds. Final programme. September 16–21, 1985. Bayreuth, West Germany.

Poiger H, Weber H, Schlatter C (1982) Special aspects of metabolism and kinetics of TCDD in dogs and rats. Assessment of toxicity of TCDD metabolite(s) in guinea pigs. *In*: Hutzinger O, Frei RW, Merian E, Pocchiari F (eds) Chlorinated dioxins and related compounds. Impact on the environment. Pergamon Press, NY. pp 317–325.

Poland A, Glover E (1979) An estimate of the maximum *in vivo* covalent binding of 2,3,7,8-tetrachlorodibenzo-*p*-dioxin to rat liver protein, ribosomal RNA and DNA. Cancer Res 39(9):3341–3344.

Poland A, Glover E (1980) 2,3,7,8-Tetrachlorodibenzo-*p*-dioxin: segregation of toxicity with the Ah locus. Mol Pharmacol 17:86–94.

Rose JQ, Ramsey JC, Wentzler TH, Hummel RA, Gehring PJ (1976) The fate of 2,3,7,8-tetrachlorodibenzo-*p*-dioxin following single and repeated oral doses to the rat. Toxicol Appl Pharmacol 36(2):209–226.

Rowe VK (1968) Kolgman/Rowe correspondence, Exhibits 865 and 866 before the Environmental Protection Agency of the United States of America.

Ryan JJ, Lizotte R, Lau B P-Y (1985) Chlorinated dibenzo-*p*-dioxins and chlorinated dibenzofurans in Canadian human adipose tissue. Chemosphere 14:697–706.

Sawahata T, Olson JR, Neal RA (1982) Identification of metabolites 2,3,7,8-tetrachlorodibenzo-*p*-dioxin (TCDD) formed on incubation with isolated rat hepatocytes. Biochem Biophys Res Commun 105(1):341–346.

Schantz SL, Barsotti DA, Allen JR (1979) Toxicological effects produced in non-human primates chronically exposed to 50 parts per trillion 2,3,7,8-tetrachlorodibenzo-*p*-dioxin (TCDD). Toxicol Appl Pharmacol 48:A180.

Schroy JM, Hileman FD, Cheng SC (1985) Physical/chemical properties of 2,3,7,8-TCDD. Chemosphere 14:877–880.

Schwetz BA, Norris JM, Sparschu GL, Rowe VK, Gehring PJ, Emerson JL, Gehring CJ (1973) Toxicology of chlorinated dibenzo-*p*-dioxins. Environ Hlth Perspect 5:87–99.

Seiler JP (1973) A survey on the mutagenicity of various pesticides. Experientia 29:622–623.

Stevens KM (1981) Agent Orange toxicity: a quantitative perspective. Human Toxicol 1:31–39.

Toth K, Sugar J, Somfai-Relle S, Bence J (1978) Carcinogenic bioassay of the herbicide, 2,4,5-trichlorophenoxyethanol (TCPE) with different 2,3,7,8-tetrachlorodibenzo-*p*-dioxin (dioxin) content in Swiss mice. Prog Biochem Pharmacol 14:82–93.

Toth K, Somfai-Relle S, Sugar J, Bence J (1979) Carcinogenicity testing of herbicide 2,4,5-trichlorophenoxyethanol containing dioxin and of pure dioxin in Swiss mice. Nature 278(5704): 548–549.

Turner JN, Collins DN (1983) Liver morphology in guinea pigs administered either pyrolysis products of a polychlorinated biphenyl transformer fluid or 2,3,7,8-tetrachlorodibenzo-*p*-dioxins. Toxicol Appl Pharmacol 67: 417–429.

USEPA[a] (1984a) Miscellaneous synthetic organic chemicals, occurrence in drinking water, food, and air. Report no. EPA-600/8-84-014A. USEPA Office of Drinking Water, Washington, DC.

USEPA[a] (1984b) Ambient water quality criteria for 2,3,7,8-tetrachlorodibenzo-*p*-dioxin. Report no. EPA-440/5-84-007. USEPA Office of Water Regulations and Standards, Washington, DC.

USEPA (1984c) Method 613. 2,3,7,8-Tetrachlorodibenzo-*p*-dioxin. Fed Regis 49(209): 433368–78. October 26, 1984.

USEPA[a] (1985a) Drinking water criteria document for 2,3,7,8-tetrachlorodibenzo-*p*-dioxin. Report no. EPA-600/X-84-194-1. USEPA Office of Drinking Water, Washington, DC.

USEPA[a] (1985b) Health assessment document for polychlorinated dibenzo-*p*-dioxins. Report no. EPA/600/8-84/014F. USEPA Environmental Criteria and Assessment Office, Cincinnati, OH.

USEPA (1986) Guidelines for carcinogenic risk assessment. Fed Regis 51(185): 33992–34003. September 24.

USFDA (1983) Food and Drug Administration. Statement by S.A. Miller, Director, Bureau of Foods, U.S. FDA, before the Subcommittee on Natural Resources, Agriculture Research and Environment, U.S. House of Representatives. June 30.

Van Miller JP, Marlar RJ, Allen JR (1976) Tissue distribution and excretion of tritiated tetrachlorodibenzo-*p*-dioxin in nonhuman primates and rats. Food Cosmet Toxicol 14(1): 31–34.

Vinopal JH, Casida JE (1973) Metabolic stability of 2,3,7,8-tetrachlorodibenzo-*p*-dioxin in mammalian liver microsomal systems and in living mice. Arch Environ Contam Toxicol 1(2): 122–132.

Vos JG, Moore JA, Zinkl JG (1974) Toxicity of 2,3,7,8-tetrachlorodibenzo-*p*-dioxin (TCDD) in C57B1/6 mice. Toxicol Appl Pharmacol 29: 299–241.

Wassom JS, Huff JE, Loprieno N (1978) A review of the genetic toxicology of chlorinated dibenzo-*p*-dioxins. Mutat Res 47(3,4): 141–160.

Young AL, Kang HK, Shepard BM (1983) Chlorinated dioxins as herbicide contaminants. Environ Sci Technol 17: 530A–540A.

Vinyl Chloride[a]

I. General Information

Vinyl chloride is also known as monochloroethylene and chloroethene. Polyvinyl chloride (PVC) is a polymer of vinyl chloride. Vinyl chloride and PVC are used as raw materials in the plastics, rubber, paper, glass, and automotive industries. They are also used in the manufacture of electrical wire insulation and cables, piping, industrial and household equipment, medical supplies, food packaging materials, and building and construction products. Vinyl chloride copolymers and PVC are distributed and processed in a variety of forms, including dry resins, plastisol (dispersions in plasticizers), organosol (dispersions in plasticizers and volatile solvent), and latex (a colloidal dispersion in water used to coat paper, fabric, or leather) (USEPA 1985a). Its structure and properties are presented in Table 1.

Vinyl chloride is a synthetic chemical with no natural sources. Since 1979, yearly U.S. production of vinyl chloride has been approximately 7 billion lb (3.18 billion kg) (USITC 1983). Vinyl chloride is polymerized, and little is released to the environment. Environmental releases are limited to the areas where vinyl chloride is produced and used.

Vinyl chloride released to air is degraded in a few hours (USEPA 1980a). Released to surface waters, it migrates to the atmosphere in a few hours or days, where it undergoes photochemical oxidation. Any of the chemical that is released to the ground does not adsorb onto soil and migrates readily to ground water. Evidence from laboratory studies suggests that vinyl chloride in ground water may degrade to CO_2 and Cl^- (Vogel and McCarty 1985). It is expected to remain in ground water for months to years. The chemical has been reported to be a degradation product of trichloroethylene and tetrachloroethylene in ground water (Parsons 1984). It does not bioaccumulate in individual animals or food chains.

Vinyl chloride does not occur widely in the environment, because of its rapid degradation and limited release. It is a relatively rare contaminant in ground and surface waters, with higher levels found in ground water. The Ground Water Supply Survey (USEPA 1982) of drinking water supplies

[a] Based on USEPA (1985a), except for Section I which is based on USEPA (1980a)

Reviews of Environmental Contamination and Toxicology, Vol 107

Table 1. General information and properties of vinyl chloride

Property	Value
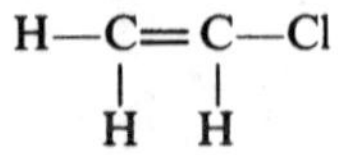	
Chemical Abstracts Service (CAS) number	75-01-4
Chemical formula	C_2H_3Cl
Molecular weight	62.5
Physical state	Gas
Boiling point	−13.3°C
Melting point	—
Density	0.9
Vapor pressure	2,530 mm Hg (20°C)
Specific gravity	0.91
Water solubility	1.1 g/L (28°C)
Log octanol/water partition coefficient	—
Odor threshold (water)	3.4 mg/L[a]
Taste threshold (water)	—
Air concentration conversion factor	2.6 mg/m^3 = 1 ppm

Source: USEPA (1985a)
[a]Amoore and Hautala (1983)

found that less than 2% of all ground-water–derived public water systems contain vinyl chloride at levels of 1 μg/L or higher. Vinyl chloride almost always co-occurs with trichloroethylene. Public systems derived from surface water also have been found to contain vinyl chloride but at lower levels.

Contaminated water is the major source of exposure to vinyl chloride. No information on the levels of vinyl chloride in food has been identified. Based upon the limited uses of vinyl chloride and its chemical properties, little or no exposure from food is expected. Vinyl chloride occurs in air in urban areas and near the sites of its production and use. Atmospheric concentrations are in the ppt range (USEPA 1979).

II. Pharmacokinetics

A. Absorption

Vinyl chloride is absorbed rapidly in rats following ingestion and inhalation (Withey 1976; Duprat et al. 1977). Using statistical modeling, Withey and Collins (1976) concluded that, for rats, a total liquid intake containing 20 ppm vinyl chloride would be equivalent to an inhalation exposure of about 2 ppm for 24 hr.

B. Distribution

After either inhalation or ingestion of ^{14}C-vinyl chloride by rats, the greatest amount of ^{14}C activity was found in the liver at 72 h; smaller amounts were observed in kidney, muscle, lung, and fat (Watanabe et al. 1976a,b). Another study of inhalation exposure of rats to ^{14}C-vinyl chloride showed the highest ^{14}C activity immediately after treatment in the liver and kidney; the next highest activity was measured in spleen and brain (Bolt et al. 1976).

C. Metabolism

Bartsch and Montesano (1975) reported two possible metabolic pathways for vinyl chloride, one involving alcohol dehydrogenase, the other mixed-function oxidase. Hefner et al. (1975) concluded that the dominant pathway at lower exposure levels probably involves alcohol dehydrogenase. Chloroethylene oxide, presumably through mixed-function oxidase, may be the principal metabolite capable of alkylating intracellular macromolecules (Laib and Bolt 1977). Vinyl chloride metabolism is saturable (i.e., the administered dose may be large enough that metabolic mechanisms of detoxification are overwhelmed) (Hefner et al. 1975; Watanabe et al. 1976a; Bolt et al. 1977).

D. Excretion

In a rat study, unmetabolized vinyl chloride was exhaled in increasing amounts as dosage (by ingestion or inhalation) was increased (Watanabe et al. 1976a,b).

Vinyl chloride metabolites are excreted primarily in the urine. In rats, urinary metabolites include *N*-acetyl-*S*-(2-hydroxyethylcysteine) and thiodiglycolic acid (Watanabe et al. 1976a).

III. Health Effects

A. Humans

At high inhalation exposure levels of 40 to 900 ppm (104 to 2,344 mg/m^3), workers have experienced dizziness, headaches, euphoria, and narcosis (USEPA 1985a). Clinical signs of chronic inhalation exposure of workers in the vinyl chloride-PVC industry include hepatotoxicity (Marsteller et al. 1975), acro-osteolysis (Lilis et al. 1975), central nervous system disturbances, pulmonary insufficiency, and cardiovascular and gastrointestinal toxicity (Miller et al. 1975; Selikoff and Hammond 1975; Suciu et al. 1975). Data on dose response in humans are scarce because few measurements of ambient vinyl chloride levels in the workplace were made before 1975 (Mancuso 1975).

Increases in the occurrence of liver angiosarcomas as well as in tumors of the brain, lung, and hematopoietic and lymphopoietic tissues have been associated with occupational exposure in workers within the vinyl chloride-

PVC industry (IARC 1979). The initial report of a link between vinyl chloride exposure and cancer (Creech and Johnson 1974), as well as subsequent reports by others, indicates the high risk and specificity of association with liver angiosarcoma, a very rare tumor.

Data on chromosomal effects of vinyl chloride exposure in workers is conflicting; both positive (Ducatman et al. 1975; Purchase et al. 1975) and negative (Killian et al. 1975; Picciano et al. 1977) results have been reported. Picciano et al. (1977) reported exposures of 0.13 to 15.2 ppm (0.34 to 40 mg/m^3, time-weighted averages) for 1 to 332 mon.

Infante et al. (1976a,b) reported an association between exposure to vinyl chloride and birth defects and fetal loss, but this association was contradicted by Edmonds et al. (1975) and Hatch et al. (1981).

B. Animals

1. Short-Term Exposure. Inhalation exposure to high levels (approximately 100,000 ppm or 260,417 mg/m^3) of vinyl chloride can induce narcosis and death, and, at lower doses, ataxia, narcosis, congestion, and edema in lungs and hyperemia in liver in several species (USEPA 1985a).

2. Long-Term Exposure. Feron et al. (1975) reported a subchronic toxicity study in which vinyl chloride monomer (VCM) dissolved in soybean oil was administered by gavage to male and female Wistar rats, initially weighing 44 g, at doses of 30, 100, or 300 mg/kg once daily, 6 d/wk for 13 wk. Several hematological, biochemical and organ weight values were significantly ($p < 0.05$) different in both mid- and high-dose animals compared to data from controls.

Inhalation exposure of rats, guinea pigs, rabbits, and dogs to 50 ppm (130 mg/m^3) vinyl chloride, 7 hr/d, 130 exposures in 189 d, did not induce toxicity, as judged by appearance, mortality, growth, hematology, liver weight, and pathology. Rats exposed to 100 ppm (260 mg/m^3) 2 hr/d for 6 mon had increased liver weights (Torkelson et al. 1961).

3. Reproductive Effects. Potential effects on reproductive capacity were not located in the literature searched.

4. Developmental Effects. Inhalation exposure of rats and rabbits to vinyl chloride concentrations as high as 2,500 ppm (6,500 mg/m^3) on d 6 to 15 (rats) and 6 to 18 (rabbits) of gestation and exposure of mice to vinyl chloride levels as high as 500 ppm (1,300 mg/m^3) on d 6 to 15 of gestation did not induce teratogenic effects but did increase skeletal variants in high-dose mice (John et al. 1977).

A developmental effects study in rats exposed by inhalation to 600 or 6,000 ppm (2,160 or 21,160 mg/m^3) 4 hr daily on gestation d 9 through 21 was negative for teratogenicity and inconclusive for fetotoxicity (Radike et al. 1977).

5. Mutagenicity. Vinyl chloride is mutagenic, presumably through active metabolites in various systems including metabolically activated systems with *Salmonella typhimurium* (Bartsch et al. 1975); *Escherichia coli* (Greim et al. 1975); yeast (Loprieno et al. 1977); germ cells of *Drosophila* (Verburgt and Vogel 1977); and Chinese hamster V79 cells (Huberman et al. 1975).

Dominant lethal studies with vinyl chloride in CD-1 mice were negative (Anderson et al. 1976).

6. Carcinogenicity. Ingestion of vinyl chloride monomer in the diet by rats at feeding levels as low as 1.7 and 5 mg/kg/d over their lifespan induced hepatocellular carcinomas and liver angiosarcomas, respectively, as well as other adverse hepatic effects (Feron et al. 1981). Til et al. (1983) extended the work of Feron et al. (1981) to include lower doses and did not find a significant ($p < 0.05$) increase in carcinogenic effects at feeding levels as high as 0.13 mg/kg/d. Administration of vinyl chloride monomer by gastric intubation for at least 52 wk resulted in carcinogenic effects in liver and other tissue sites in rats (Feron et al. 1981; Maltoni et al. 1981).

Chronic inhalation of vinyl chloride has induced cancer in liver and other tissue sites in rats and mice (Lee et al. 1977, 1978; Maltoni et al. 1981).

IV. Quantification of Toxicological Effects[a]

A. One-Day Health Advisory

There are insufficient data for calculation of a One-Day Health Advisory. The Ten-Day HA of 2.6 mg/L is proposed as a conservative estimate of the One-Day HA.

B. Ten-Day Health Advisory

The inhalation data by Torkelson et al. (1961) were not selected for the Ten-Day HA calculation because studies of oral exposure are more relevant to a health advisory for drinking water. Instead, the Feron et al. (1975) study on rats was chosen. Using the No-Observed-Adverse-Effect Level (NOAEL) of 30 mg/kg from this study, the Ten-Day HA for a 10-kg child is calculated as follows:

$$\text{Ten-Day HA} = \frac{(30\ \text{mg/kg/d})(6/7)(10\ \text{kg})}{(100)(1\ \text{L/d})} = 2.6\ \text{mg/L}\ (2{,}600\ \mu\text{g/L})$$

[a] See Introduction for discussion of methodology

where 6/7 = conversion of 6 d/wk treatment to 7 d/wk to represent daily exposure.

C. Longer-Term Health Advisory

The Longer-Term HA can be calculated using data from the lifetime feeding study in rats by Til et al. (1983). Carcinogenic and noncarcinogenic effects in the liver were evident with a vinyl chloride dietary level of 1.3 mg/kg/d. At dietary levels of 0.014 and 0.13 mg/kg/d, increased incidences of basophilic foci of cellular alteration in the liver of female rats were evident. However, basophilic foci by themselves do not represent an adverse effect on the liver in the absence of evidence of liver lesions, such as those found in the 1.3 mg/kg/d group; and a dose-related increase in basophilic foci was not evident. Therefore, the dose of 0.13 mg/kg/d is identified as the NOAEL for noncarcinogenic effects for the Longer-Term HA calculation.

The Longer-Term HA for a 10-kg child is calculated as follows:

$$\text{Longer-Term HA} = \frac{(0.13 \text{ mg/kg/d})(10 \text{ kg})}{(100)(1 \text{ L/d})} = 0.013 \text{ mg/L } (13\ \mu\text{g/L})$$

The Longer-Term HA for a 70-kg adult is calculated as follows:

$$\text{Longer-Term HA} = \frac{(0.13 \text{ mg/kg/d})(70 \text{ kg})}{(100)(2 \text{ L/d})} = 0.046 \text{ mg/L } (46\ \mu\text{g/L})$$

D. Lifetime Health Advisory

Because vinyl chloride is classified as a human carcinogen (IARC Group 1 and USEPA Group A), a Lifetime Health Advisory is not recommended.

E. Evaluation of Carcinogenic Potential

With the criteria described in USEPA's guidelines for assessment of carcinogenic risk (USEPA 1986), vinyl chloride may be classified in Group A: human carcinogen. This category is for agents for which there is sufficient evidence to support the causal association between exposure to the agents and cancer. The International Agency for Research on Cancer (IARC 1979) has concluded that there is sufficient evidence to classify vinyl chloride as a human carcinogen in Category 1.

USEPA's Carcinogen Assessment Group (CAG) has recalculated its excess carcinogenic risk estimates resulting from lifetime exposure to vinyl chloride through drinking water (USEPA 1985a). CAG based its preliminary revised estimates on the Feron et al. (1981) study. The total number of tumors of lung and liver in rats exposed through diet was used to calculate the excess cancer risk. Using the 95% upper limit [$q_1{}^* = 2.3\ (\text{mg/kg/d})^{-1}$] with the linearized multistage model, CAG calculated that consuming 2 L of water per day with vinyl chloride concentrations of 1.5 μg/L, 0.15 μg/L, or 0.015 μg/L would increase the risk of one excess cancer per 10,000 (10^{-4}), 100,000 (10^{-5}), or 1,000,000 (10^{-6}) people exposed, respectively, per lifetime. The CAG is presently reassessing the cancer risk estimate based on the study by Feron et al. (1981) by taking into account the more recent study by Til et al. (1983) which includes lower doses than those studied earlier.

Examples of Maximum Likelihood Estimates (MLE) as well as 95% upper limits (UL) of cancer risks by the multistage model are as follows (risk is expressed as cases/lifetime/person): 0.01 mg/kg/d or 0.35 mg/L exposure is associated with risks of 1.6×10^{-2} (MLE) and 1.9×10^{-2} (UL) and 0.001 mg/kg/d exposure is associated with risks of 1.6×10^{-3} (MLE) and 1.9×10^{-3} (UL).

Cancer risk estimates (95% upper limit) from other models can be compared to those derived using the multistage model. For example, one excess cancer per 1,000,000 (10^{-6}) population is associated with exposure to vinyl chloride in drinking water at levels of 50 μg/L (probit), 0.5 μg/L (logit), and 0.02 μg/L (Weibull). While these are recognized as statistically alternative approaches, the range of risks described by using any of these modeling approaches has little biological significance unless the data support the selection of one model over another. In the interest of consistency of approach and in providing an upper-bound estimate of the potential cancer risk, USEPA has recommended use of the linearized multistage approach.

V. Other Criteria, Guidance, and Standards

The National Academy of Sciences (NAS 1977) estimated a 10^{-6} risk (95% upper-bound estimate) from lifetime exposure to 1 μg/L drinking water with the multistage model and the lifetime ingestion study in rats by Maltoni et al. (1981).

The final Recommended Maximum Contaminant Level (RMCL) by the USEPA Office of Drinking Water is zero (USEPA 1985b); the final MCL is 2 μg/L (USEPA 1987).

Ambient water quality criteria (USEPA 1980b) are 20, 2, and 0.2 μg/L for risks of 10^{-5}, 10^{-6}, and 10^{-7}, respectively, assuming consumption of 2 L of water and 6.5 g of contaminated fish per day by a 70-kg adult.

A workplace standard of 1 ppm (Time-Weighted Average) was set by the Office of Safety and Health Administration in 1974 based on the demonstration of angiosarcoma of the liver in vinyl chloride workers (OSHA 1974).

The American Conference of Governmental Industrial Hygienists (ACGIH 1982) has recommended a Threshold Limit Value of 5 ppm (10 mg/m^3).

VI. Analytical Methods

Analysis of vinyl chloride levels is done by a purge-and-trap gas chromatographic procedure used for the determination of volatile organohalides in drinking water (USEPA 1985c). This method calls for bubbling an inert gas through a sample of water and the trapping of purged vinyl chloride on an adsorbent. The adsorbent is heated to drive off the vinyl chloride onto a gas chromatographic column. This method can be used to measure vinyl chloride over a concentration range of 0.06 to 1,500 μg/L. Confirmatory analysis for vinyl chloride is by mass spectrometry (USEPA 1985d). The detection limit for confirmation by mass spectrometry is 0.3 μg/L.

VII. Treatment Technologies

The value of Henry's Law constant for vinyl chloride (6.4 atm-m^3/mole) suggests aeration is a potential removal technique for vinyl chloride in water (ESE 1984). Removal efficiencies up to 99.27% were achieved at 9°C using a pilot packed tower aerator. In similar studies, vinyl chloride was removed from ground water using a spray aeration system, with total volatile organic chemical (VOC) concentration 100 to 200 μg/L (ESE 1984). Greater than 99.9% VOC removal was obtained using a four-stage aeration system; each stage employed 20 shower heads with a pressure drop of approximately 10 lb/in^2 (704 g/cm^2). In-well aeration with an air-life pump has also demonstrated up to 97% removal of vinyl chloride. However, practical considerations are likely to limit the application of this technique (Miltner 1984).

The concentration of vinyl chloride in southern Florida ground water decreased by 25 to 52% following passage through lime-softening basins and filters (Wood and DeMarco 1980). Since vinyl chloride is a highly volatile compound, it is probably volatilized during treatment (ESE 1984).

Adsorption techniques have been less successful than aeration in removing vinyl chloride from water. In a pilot study, water from a ground-water treatment plant was passed through a series of four 30-in (76.2-cm) granular-activated carbon (Filtrasorb 400®) columns (Wood and DeMarco 1980; Symons 1978); the empty bed contact time was approximately 6 min per column. Influent vinyl chloride concentrations ranged from below detection to 19 μg/L; erratic removal was reported. To maintain effluent concentrations below 0.5 μg/L, the estimated column capacity to breakthrough was 810, 1,250, 2,760, and 2,050 bed volumes for empty bed contact times of 6, 12, 19, and 25 min, respectively. In addition, the estimated service life of the activated carbon was low. Similarly, poor removal of vinyl chloride was achieved using an experimental synthetic resin, Ambersorb XE-340® (Symons 1978).

Treatment technologies for the removal of vinyl chloride from water have not been extensively evaluated except on an experimental level. The selection of individual or combined technologies to achieve vinyl chloride removal must be based on a case-by-case technical evaluation as well as an assessment of the economics involved.

References

ACGIH (1982) American Conference of Governmental Industrial Hygienists. TLV's —Threshold limit values for chemical substances and physical agents in the work-room environment. ACGIH, Cincinnati, OH.

Amoore JE, Hautala E (1983) Odor as an aid to chemical safety: Odor thresholds compared with threshold limit values and volatilities for 214 industrial chemicals in air and water dilution. J Appl Toxicol 3:272–290.

Anderson D, Hodge MCE, Purchase IFH (1976) Vinyl chloride: dominant lethal studies in male CD-1 mice. Mutat Res 40:359–370.

Bartsch H, Malaveille C, Montesano R (1975) Human, rat and mouse liver-mediated mutagenicity of vinyl chloride in *S. typhimurium* strains. Int J Cancer 15:429–437.

Bartsch H, Montesano R (1975) Mutagenic and carcinogenic effects of vinyl chloride. Mutat Res 32:93–114.

Bolt HM, Kappus H, Buchter A, Bolt W (1976) Disposition of (1,2,-^{14}C) vinyl chloride in the rat. Arch Toxicol 35:153–162.

Bolt HM, Laib RJ, Kappus H, Buchter A (1977) Pharmacokinetics of vinyl chloride in the rat. Toxicology 7:179–188.

Creech JL, Johnson MN (1974) Angiosarcoma of the liver in the manufacture of polyvinyl chloride. J Occup Med 20:338–340.

Ducatman A, Hirschhorn K, Selikoff IJ (1975) Vinyl chloride exposure and human chromosome aberrations. Mutat Res 31:163–168.

Duprat P, Fabry JP, Gradiski D, Magadur JL (1977) Metabolic approach to industrial poisoning: blood kinetics and distribution of ^{14}C-vinyl chloride monomer (V.C.M.). Acta Pharmacol Toxicol Suppl (Kbh) 41(1):142–143.

Edmonds LD, Falk H, Nissim JE (1975) Congenital malformations and vinyl chloride. Lancet 2:1098.

ESE[a] (1984) Environmental Science and Engineering. Technologies and costs for the removal of volatile organic chemicals from potable water supplies. (Draft) ESE No. 84-912-0300. Prepared for USEPA Science and Technology Branch, Criteria and Standards Division (CSD), Office of Drinking Water, Washington, DC.

Feron VJ, Speek AJ, Williams MI, van Battum D, de Groot AF (1975) Observations on the oral administration and toxicity of vinyl chloride in rats. Food Cosmet Toxicol 13:633–638.

Feron VJ, Hendrikson CFM, Speek AJ, Til HP, Spit BJ (1981) Lifespan oral toxicity study of vinyl chloride in rats. Food Cosmet Toxicol 19:317–331.

Greim H, Bonse G, Radwan Z, Reichert D, Henschler D (1975) Mutagenicity *in vitro*

[a] USEPA documents can be obtained by writing to the appropriate office within the U.S Environmental Protection Agency, 401 M Street, S W., Washington, DC 20460. They may also be available by contacting the National Technical Information Service, 5285 Port Royal Road, Springfield, VA 22161 (Telephone 1-800-336-4700)

and potential carcinogenicity of chlorinated ethylenes as a function of metabolic oxirane formation. Biochem Pharmacol 24:2013–2017.

Hatch M, Kline J, Stain Z (1981) Power considerations in studies of reproductive effects of vinyl chloride and some structural analogs. Environ Hlth Perspec 41: 195–201.

Hefner RE Jr, Watanabe PG, Gehring PJ (1975) Preliminary studies on the fate of inhaled vinyl chloride monomer in rats. Ann NY Acad Sci 246:135–148.

Huberman E, Bartsch H, Sachs L (1975) Mutation induction in Chinese hamster V79 cells by two vinyl chloride metabolites, chloroethylene oxide and 2-chloroacetaldehyde. Int J Cancer 16:369–644.

IARC (1979) International Agency for Research on Cancer. IARC monographs on the evaluation of carcinogenic risk of chemicals to man. Vol 19, pp 377–438. IARC, Lyon, France.

Infante PF, Wagoner JK, Waxweiler RJ (1976a) Carcinogenic, mutagenic, and teratogenic risks associated with vinyl chloride. Mutat Res 41:131–142.

Infante PF, Wagoner JK, Waxweiler RJ, McMichael AJ, Falk H (1976b) Genetic risks of vinyl chloride. Lancet 1:734–735.

John JA, Smith FA, Leong BKJ, Schwetz BA (1977) The effects of maternally inhaled vinyl chloride on embryonal and fetal development in mice, rats and rabbits. Toxicol Appl Pharmacol 39:497–513.

Killian DJ, Picciano DJ, Jacobson CB (1975) Industrial monitoring: A cytogenetic approach. Ann NY Acad Sci 269:4–11.

Laib RJ, Bolt HM (1977) Alkylation of RNA by vinyl chloride metabolites *in vitro* and *in vivo*: Formation of 1-*N'*-etheno-adenosine. Toxicology 8:185–195.

Lee CC, Bhandari JC, Winston JM, House WB, Dixon RL, Woods JS (1977) Inhalation toxicity of vinyl chloride and vinylidene chloride. Environ Hlth Perspect 21:25–32.

Lee CC, Bhandari JC, Winston JM, House WB, Dixon RL, Woods JS (1978) Carcinogenicity of vinyl chloride and vinylidene chloride. J Toxicol Environ Hlth 4: 15–30.

Lilis R, Anderson H, Nicolson WJ, Daum S, Fischbein AS, Selikoff IJ (1975) Prevalence of disease among vinyl chloride and polyvinyl chloride workers. Ann NY Acad Sci 246:22–41.

Loprieno N, Barale R, Baroncelli S, Bartsch H, Bronzetti G, Cammellini A, Corsi C, Frezza D, Nieri R, Leporini D, Rosellini D, Rossi AM (1977) Induction of gene mutations and gene conversions by vinyl chloride metabolites in yeast. Cancer Res 253–257.

Maltoni C, Lefemine G, Ciliberti A, Cotti G, Carretti D (1981) Carcinogenicity bioassays of vinyl chloride monomer: a model of risk assessment on an experimental basis. Environ Hlth Perspec 41:3–31.

Mancuso TF (1975) Comments for opening of discussion on "neoplastic effects." Ann NY Acad Sci 246:251–254.

Marsteller HJ, Lelbach WK, Muller R, Gedigk P (1975) Unusual splenomegalic liver disease as evidenced by peritoneoscopy and guided liver biopsy among polyvinyl chloride production workers. Ann NY Acad Sci 246:95–134.

Miller A, Tiersten AS, Chuang M, Selikoff IJ, Warshaw R (1975) Changes in pulmonary function in workers exposed to vinyl chloride and polyvinyl chloride. Ann NY Acad Sci 246:42–52.

Miltner R[a] (1984) Personal communication. Cited in: Technologies and costs for the removal of volatile organic chemicals from potable water supplies by environmental science and engineering. USEPA Technical Support Division of Office of Drinking Water (ODW), Cincinnati, OH.

NAS (1977) National Academy of Sciences. Drinking water and health, Vol 1. National Academy Press, Washington, DC. pp. 783–787.

OSHA (1974) Vinyl chloride: exposure standards. Fed Regis 39:3589.

Parsons F, Wood PR, DeMarco J (1984) Transformation of tetrachloroethene and trichloroethene in microcosms and groundwater, J Am Waterworks Assoc 26(2): 56f.

Picciano DJ, Flake RE, Gay PC, Killian DJ (1977) Vinyl chloride cytogenetics. J Occup Med 19:527–530.

Purchase IFH, Richardson CR, Anderson D (1975) Chromosomal and dominant lethal effects of vinyl chloride. Lancet 2(7931):410–411.

Radike M et al. (1977) Transplacental effects of vinyl chloride in rats. Annual report, pp 183–185. USPHS-ES-00159. Center for Study of the Human Environment, Department of Environmental Health, University of Cincinnati Medical Center, Cincinnati, OH.

Selikoff IJ, Hammond EC (eds) (1975) Toxicity of vinyl chloride-polyvinyl chloride. Ann NY Acad Sci, Vol 246.

Suciu I, Prodan L, Ilea E, Paduracu A, Pascu L (1975) Clinical manifestations in vinyl chloride poisoning. Ann NY Acad Sci 246:53–69.

Symons JM[a] (1978) Interim treatment guide for controlling organic contaminants in drinking water using granular activated carbon. USEPA Office of Research and Development (ORD), Municipal Environmental Research Laboratory, Cincinnati, OH. Cited in USEPA SNARL Document for vinyl chloride (draft) and in USEPA, May 1983, Treatment of volatile organic compounds in drinking water. Report no. EPA-600/8-83-019, Office of Research and Development, Municipal Environmental Research Laboratory, Drinking Water Research Division, Cincinnati, OH.

Til HP, Immel HR, Feron VJ[a] (1983) Lifespan oral carcinogenicity study of vinyl chloride in rats. Final report. Civo Institutes TNO. Report no. V 83.285/291099. Submitted to USEPA Office of Drinking Water, Washington, DC.

Torkelson RR, Oyen F, Rowe VK (1961) The toxicity of vinyl chloride as determined by repeated exposure of laboratory animals. Am Ind Hyg Assoc J 22:354–361.

USEPA[a] (1979) Water related environmental fate of 129 priority pollutants. Report no. EPA-440/4-79-029. USEPA Office of Water Planning and Standards, Washington, DC.

USEPA[a] (1980a) Vinyl chloride occurrence in drinking water, food, and air. USEPA Office of Drinking Water, Washington, DC.

USEPA[a] (1980b) Ambient water quality criteria for vinyl chloride. Report no. EPA-440/5-80-078. USEPA Office of Water Regulations and Standards, Washington, DC.

USEPA[a] (1982) The ground-water supply survey—Summary of volatile organic contaminant occurrence data. USEPA Office of Drinking Water, Technical Support Division, Cincinnati, OH. June.

USEPA[a] (1985a) Final draft for the drinking water criteria document on vinyl chloride. TR-540-162. USEPA Office of Drinking Water, Washington DC.

USEPA[a] (1985b) National primary drinking water regulations; Volatile synthetic organic chemicals, Final rule and proposed rule. Fed Regis 50(219):46880–933. November 13.

USEPA[a] (1985c) Method 502.1. Volatile halogenated organic compounds in water by purge-and-trap gas chromatography. USEPA Environmental Monitoring and Support Laboratory, Cincinnati, OH. June 1985.

USEPA[a] (1985d) Method 524.1. Volatile organic compounds in water by gas chromatography/mass spectrometry. USEPA Environmental Monitoring and Support Laboratory, Cincinnati, OH. June 1985.

USEPA[a] (1986) Guidelines for carcinogenic risk assessment. Fed Regis 51(185): 33993–34003. September 24.

USEPA[a] (1987). National primary drinking water regulations; Synthetic organic chemicals; Monitoring for unregulated contaminants; Final rule. Fed Regis 52(130):25690–25717. July 8.

USITC (1983) U.S. International Trade Commission. Synthetic organic chemicals, United States production, 1982. USITC publication 1422. USITC, Washington, DC.

Verburgt FG, Vogel E (1977) Vinyl chloride mutagenesis in *Drosophila melanogaster*. Mutat Res 48:327–33.

Vogel T, McCarty P (1985) Biotransformation of tetrachloroethylene to trichloroethylene, dichloroethylene, vinyl chloride, and carbon dioxide under methanogenic conditions. Appl Environ Microbiol 49(5):1080–83.

Watanabe PG, McGowan GR, Gehring PJ (1976a) Fate of [^{14}C]-vinyl chloride after single oral administration in rats. Toxicol Appl Pharmacol 36:339–52.

Watanabe PG, McGowan GR, Madrid EO, Gehring PJ (1976b) Fate of [^{14}C]-vinyl chloride following inhalation exposure in rats. Toxicol Appl Pharmacol 37:49–59.

Withey JR (1976) Pharmacodynamics and uptake of vinyl chloride monomer administered by various routes to rats. J Toxicol Environ Hlth 1:381–94.

Withey JR, Collins BT (1976) A statistical assessment of the quantitative uptake of vinyl chloride monomer from aqueous solution. J Toxicol Environ Hlth 2:311–21.

Wood PR, DeMarco J (1980) Effectiveness of various adsorbents in removing organic compounds from water. 1. Removing purgeable halogenated organics. In: McGuire MJ, Suffet IH (eds) Activated carbon adsorption of organics from the aqueous phase, Vol 2. Ann Arbor Science, Ann Arbor, MI. pp 85–114.

Appendix A: Health Advisories (HAs) for Inorganic Chemicals in Drinking Water (mg/L)

	One-Day HA	Ten-Day HA	Longer-term HA for 10-kg child	Longer-term HA for 70-kg adult	DWEL	DWEL adjusted for 10-kg child	Lifetime HA
Acrylamide	1.5	0.3	0.02	0.07	0.007	—[d]	—[c]
Barium	—[e]	—[e]	—[e]	—[e]	5	—[d]	5
Cadmium	0.043	0.043	—[b]	0.018	0.018	0.005	0.005
Chromium	—[a]	1.4	0.24	0.84	0.17	—[d]	0.12
Cyanide	—[b]	—[b]	0.22	0.77	0.77	0.22	0.154
Epichlorohydrin	—[a]	0.14	—[d]	0.07	0.07	—[d]	—[c]
Mercury	—[b]	—[b]	—[b]	—[b]	0.0055	0.00158	0.0011
Nickel	—[a]	1.0	—[b]	0.58	0.58	0.16	0.17
Nitrate-nitrogen (4-kg child)	—[a]	10	10	10	—[d]	—[d]	10
Nitrate-nitrogen (all others)	—[a]	111	10	10	—[d]	—[d]	10
Nitrite-nitrogen (4-kg child)	—[a]	1	1	1	—[d]	—[d]	1
Nitrite-nitrogen (all others)	—[a]	11	1	1	—[d]	—[d]	1
Styrene	22.5	—[f]	2	7	0.7	—[d]	—[c]
TCDD	1.0×10^{-6}	1.0×10^{-7}	1.0×10^{-8}	3.5×10^{-8}	3.5×10^{-8}	—[d]	—[c]
1,1,1-Trichloroethane	140	—[f]	35	125	1.0	—[d]	0.2
Vinyl chloride	—[a]	2.6	0.013	0.046	—[d]	—[d]	—[c]

[a] Data unavailable to derive One-Day HA; application of Ten-Day HA is recommended.

[b] Not determined, use of DWEL, adjusted for 10-kg child, is recommended as estimate

[c] Not recommended, probable human (Group B2) carcinogen.

[d] Not calculated.

[e] Data insufficient for calculation; use of Lifetime HA is recommended.

[f] Appropriate data unavailable for calculation; use of Longer-Term HA for 10-kg child is recommended as estimate.

Index

INFORMATION FOR AUTHORS

Reviews of Environmental Contamination and Toxicology

Edited by

George W. Ware

Published by
Springer-Verlag New York • Berlin • Heidelberg • Tokyo

The original copy and one good photocopy of the manuscript, complete with figures and tables, are required. Manuscripts will be published in the order in which they are received, reviewed, and accepted. They should be sent to the editor:

Dr. George W. Ware
College of Agriculture
University of Arizona
Tucson, Arizona 85721
Telephone: (602) 621-3859 (office)
(602) 299-3735 (home)

1. Manuscript

The manuscript, in English, should be typewritten, double-spaced throughout (including reference section), on one side of 8½ × 11-inch blank white paper, with at least one-inch margins. The first page of the manuscript should start with the title of the manuscript, name(s) of author(s), with author affiliation(s) as first-page starred footnotes, and "Contents" section. Pages should be numbered consecutively in arabic numerals, including those bearing figures and tables only. In titles, in-text outline headings and subheadings, figure legends, and table headings only the initial word, proper names, and universally capitalized words should be capitalized.

Footnotes should be inserted in text and numbered consecutively in the text using arabic numerals.

Tables should be typed on separate sheets and numbered consecutively within the text in *roman numerals*; they should bear a descriptive heading, in lower case, which is underscored with one line and which starts after the word "Table" and the appropriate roman numeral; *footnotes in tables* should be designated consecutively within a table by the lower-case alphabet. *Figures* (including photos, graphs, and line drawings) should be numbered consecutively within the text in arabic numerals; each figure should be affixed to a separate page bearing a legend (below the figure) in lower case starting with the term "Fig." and a number.

2. Summary

A concise but informative summary (double-spaced) must conclude the text of each manuscript; it should summarize the significant content and major conclusions presented. It must not be longer than two 8½ × 11-inch pages of double-spaced typing. As a summary, it should be more informative than the usual abstract.

3. References

All papers, books, and other works cited in the text must be included in a "References" section (*also double-spaced*) at the end of the manuscript. If comprehensive papers on the same subject have been published, they should be cited when the bibliographic citations extend farther back than to these papers.

All papers cited in the text should be given in parentheses and alphabetically when more than one reference is cited at a time, e.g. (Coats and Smith 1979, Holcombe et al. 1982, Stratton 1986), except when the author is mentioned, as for example, "and the study of Roberts and Stoydin (1985)." References to unpublished works should be kept to a minimum and mentioned only in the text itself in parentheses. References to published works are given at the end of the text in alphabetical order under the first author's name, citing all authors (surnames followed by initials throughout; do not use "and") according to the following examples:

Periodicals: Name(s), initials, year of publication in parentheses, full article title, journal title as abbreviated in "The ACS Style Guide: A Manual for Authors and Editors" of the American Chemical Society, volume number, colon, first and last page numbers. Example:

Leistra MT (1970) Distribution of 1,3-dichloropropene over the phases in soil. J Agric Food Chem 18:1124–1126.

Books: Name(s), initials, year of publication in parentheses, full title, edition, volume number, name of publisher, place of publication, first and last page numbers. Example:

Gosselin R, Hodge H, Smith R, Gleason M (1976) Clinical Toxicology of Commercial Products, 4th Ed. Wilkins-Williams, Baltimore, MD, pp 119–121.

Work in an edited collection: Name(s), initials, year of publication in parentheses, full title. In: name(s) and initial(s) of editor(s), the abbreviation ed(s) in parentheses, name of publisher, place of publication, first and last page numbers. Example:

Metcalf RL (1978) Fumigants. In: White-Stevens J (ed) Pesticides in the environment. Marcel Dekker, New York. pp 120–130.

References by the same author(s) are arranged chronologically. If more than one reference by the same author(s) published in the same year is cited, use a, b, c after year of publication in both text and reference list.

4. Illustrations

Illustrations may be included only when indispensable for the comprehension of text. They should not be used in place of concise explanations in text. Schematic line drawings must be drawn carefully. For other illustrations, clearly defined black-and-white glossy photos are required. Should darts (arrows) or letters be required on a photo or other type of illustration, they should be marked neatly with a soft pencil on a duplicate copy or on an overlay, with the end of each dart indicated by a fine pinprick; darts and lettering will be transferred to the illustrations by the publisher.

Photos should not be less than five × seven inches in size. Alterations of photos in page proof stage are not permitted. *Each photograph or other illustration should be marked on the back, distinctly but lightly, with soft pencil, with first author's name, figure number, manuscript page number, and the side which is the top.*

If illustrations from published books or periodicals are used, the exact source of each should be included in the figure legend: if these "borrowed" illustrations are copyrighted by others, permission of the copyright holder to reproduce the illustration must be secured by the author.

5. Chemical Nomenclature

All pesticides and other subject-matter chemicals should be identified according to *Chemical Abstracts*, with the full chemical name in text in parentheses or brackets the first time a common or trade name is used. *If many such names are used, a table of the names, their precise chemical designations, and their* Chemical Abstract Numbers (CAS) *should be included as the last table in the manuscript, with a numbered footnote reference to this fact on the first text page of the manuscript.*

6. Miscellaneous

Abbreviations: Common units of measurement and other commonly abbreviated terms and designations should be abbreviated as listed below; if any others are used often in a manuscript, they should be written out the first time used, followed by the normal and acceptable abbreviation in parentheses [e.g., Acceptable Daily Intake (ADI), Angstrom (Å), picogram (pg)]. Except for inch (in.) and number (no., when followed by a numeral), abbreviations are used without periods. Temperatures should be reported as "°C" or "°F" (e.g., mp 41° to 43°C). Because the metric system is the international standard, when pounds (lb) and gallons (gal) are used the metric equivalent should follow in parentheses.

Abbreviations

A	acre	sec	second(s)
bp	boiling point	μg	microgram(s)
cal	calorie	μL	microliter(s)
cm	centimeter(s)	μm	micrometer(s)
cu	cubic (as in "cu m")	mg	milligram(s)
d	day	mL	milliliter(s)
ft	foot (feet)	mm	millimeter(s)
gal	gallon(s)	mM	millimolar
g	gram(s)	min	minute(s)
ha	hectare	M	molar
hr	hour(s)	mon	month(s)
in.	inch(es)	ng	nanogram(s)
id	inside diameter	nm	nanometer(s) (millimicron)
kg	kilogram(s)	N	normal
L	liter(s)	no.	number(s)
mp	melting point	od	outside diameter
m	meter(s)	oz	ounce(s)
ppb	parts per billion	sp gr	specific gravity
ppm	parts per million	sq	square (as in "sq m")
ppt	parts per trillion	vs	versus
pg	picogram	wk	week(s)
lb	pound(s)	wt	weight
psi	pounds per square inch	yr	year(s)
rpm	revolutions per minute		

Numbers: All numbers used with abbreviations and fractions or decimals are arabic numerals. Table numbers are roman numerals. Otherwise, numbers below ten are to be written out. Numerals should be used for a series (e.g., "0.5, 1, 5, 10, and 20 days"), for pH values, and for temperatures. When a sentence begins with a number, write it out.

Symbols: Special symbols (e.g., Greek letters) must be identified in the margin, e.g.

$$A = \beta/2\lambda$$

beta (β); lambda (λ)

Percent should be % in text, figures, and tables.

Style and format: The following examples illustrate the style and format to be followed (except for abandonment of periods with abbreviation):

Sklarew DS, Girvin DC (1986) Attenuation of polychlorinated biphenyls in soils. Reviews Environ Contam Toxicol 98:1–41.

Yang RHS (1986) The toxicology of methyl ethyl ketone. Residue Reviews 97:19–35.

7. Proofreading scheme

The senior author must return the Master set of page proof to the Editor within one week of receipt. Author corrections should be clearly indicated on proof with ink, and in conformity with the standard "Proofreader's Marks" accompanying each set of proofs. In correcting proof, new or changed words or phrases should be carefully and legibly handprinted (not handwritten) in the margins.

8. Offprints

Senior authors receive 30 complimentary offprints of a published article. Additional offprints may be ordered from the publisher at the time the principal author receives the proof. Order forms for additional offprints will be sent to the senior author along with the page proofs.

9. Page charges

There are no page charges, regardless of length of manuscript. However, the cost of alteration (other than corrections of typesetting errors) attributable to authors' changes in the page proof, in excess of 10% of the original composition cost, will be charged to the authors.

If there are further questions, see any volume of *Reviews of Environmental Contamination and Toxicology* (formerly *Residue Reviews*) or telephone the Editor (see first page for telephone numbers). Volume 98 is especially helpful for style and format.

GPSR Compliance
The European Union's (EU) General Product Safety Regulation (GPSR) is a set of rules that requires consumer products to be safe and our obligations to ensure this.

If you have any concerns about our products, you can contact us on

ProductSafety@springernature.com

In case Publisher is established outside the EU, the EU authorized representative is:

Springer Nature Customer Service Center GmbH
Europaplatz 3
69115 Heidelberg, Germany

www.ingramcontent.com/pod-product-compliance
Lightning Source LLC
LaVergne TN
LVHW052323200926
848481LV00002B/651
* 9 7 8 1 4 6 8 4 7 0 8 5 7 *